SKINNY GENES

THE SURPRISING TRUTH ABOUT EVERY BODY'S CAPACITY TO SETTLE AT A NATURAL WEIGHT, EVEN WHEN DIETS HAVE FAILED

ARIANNE BOZARTH, CNS, MS

Difference Press

Washington, DC, USA

Copyright © Arianne Bozarth 2021

All rights reserved. No part of this book may be reproduced in any form without permission in writing from the author. Reviewers may quote brief passages in reviews.

Published 2021

DISCLAIMER

All content contained in this book, including, but not limited to, text, images, and information herein or in any other formats of the book (the "Content") was created for informational purposes only. The Content is not intended to be a substitute for professional medical advice, diagnosis, or treatment. The Content is not intended to diagnose, treat, cure, or prevent any disease or medical condition. The Content has not been evaluated by the Food and Drug Administration. Please consult your physician or healthcare provider if you have any medical condition or any medical questions or concerns. NEVER DISREGARD PROFESSIONAL MEDICAL ADVICE OR DELAY IN SEEKING IT BECAUSE OF SOMETHING YOU HAVE READ IN THIS BOOK. Aceso Nutrition LLC, the individual author, and the publisher of this book shall not be held liable for any damages, including, but not limited to, special, incidental, consequential, or other damages.

No part of this publication may be reproduced or transmitted in any form or by any means, mechanical or electronic, including photocopying or recording, or by any information storage and retrieval system, or transmitted by email without permission in writing from the author.

Neither the author nor the publisher assumes any responsibility for errors, omissions, or contrary interpretations of the subject matter herein. Any perceived slight of any individual or organization is purely unintentional.

Brand and product names are trademarks or registered trademarks of their respective owners.

Cover design: Jennifer Stimson

Editing Natasa Smirnov

Author's photo courtesy of: Stacey Vaeth of Stacey Vaeth Photography

CONTENTS

1. Why can't I lose weight? — 1
2. Uncovering the Skinny Gene Designer's Blueprint — 5
3. Where Genes Meet Design — 11
4. Starting with the "Food of Gods" — 19
5. Getting into the Flow — 33
6. Priming for the Burn — 49
7. Sparking the Fire — 61
8. Igniting the Fire — 77
9. Fanning the Flame — 95
10. Fueling the Fire — 127
11. On the Way to Skinny Me — 141

References — 147
About the Author — 159
About Difference Press — 161
Other Books by Difference Press — 165
Thank You — 167

To my greatest loves, Kyle, Nathaniel, and Sophia, who daily light my flame and spark my never-ending curiosity and awe. I owe my expansion to you ... love you forever and always!

1

WHY CAN'T I LOSE WEIGHT?

Gretchen puts down the knife on the cutting board and pulls up her jeans. She likes the cut, but they are not deep enough to contain her double post-partum belly. Ever since she had her two babies, she's been struggling to achieve her weight goals. That's probably slightly bending the truth. In reality, she's been struggling for most of her life. She stirs the veggies and adds some seasoning, already dreading the negotiations that are sure to ensue with her kiddos at the dinner table. But she is determined to serve a healthy meal. Well, at least that's what the last book defined as healthy. She loves to cook and is always trying new recipes based on the latest and greatest health foods, so she can support her family in eating better and staying healthy. Her husband teases her about changing the weekly menu on them with every new "health solution" she uncovers, but it's really more than that. She is just trying to figure out what works in keeping herself and her family healthy. *I have a PhD, I manage a team of fifty, I should really be able to figure this out. How do I get over this wall I've hit?*

Despite putting in all her best efforts and energy to learn what she should be eating, buying only the best, high-quality ingredients at the health food store, and spending time creating recipes that taste good for the whole family, she is still at the same weight – about fifteen pounds above her ideal weight. Gretchen feels stuck; she feels like she is doing all the right things that she is supposed to be doing, but she is not seeing the results that she wants. She can't figure out why her body won't respond. According to all she has read, the way she eats and all her lifestyle choices should have her at her ideal weight by now. Over the last three months, her weight has even started sliding back up a bit and she feels like her body is working against her. Whenever she thinks about it, it makes her incredibly frustrated. Sometimes she makes excuses that it is just her body getting older, but deep down she feels like that isn't it. Then she starts to worry that maybe something is wrong with her body. What if her recent inability to lose weight is a sign of greater health issues to come? What if she can never lose the weight? Thinking about staying at this current weight indefinitely into the future, or even worse – gaining more weight again and not being able to do anything about it

– makes her feel like she is no longer in control of her body. Her biggest question is: What am I doing wrong?

If you picked up this book, you've probably spent a decent amount of time trying different ways of eating and exercising. You've been riding the emotionally and mentally draining rollercoaster called "Calories In – Calories Out" for quite some time now. Maybe some ways were more successful than others and some may have brought you results … for a while. However, here you are with another book in your hands hoping for a long-lasting solution. The question is, what is the problem you are actually solving? Why did some previous diets work and some didn't? Why haven't you been able to find one that fits for a lifetime? Short answer: Because while you were trying to get into your old skinny jeans, no one told you about the umpteen beneficial effects of unlocking your skinny genes (pun intended).

If Gretchen's story resonates with you and if you are ready to find out what is holding you back from rediscovering that "perfect fit" diet to unlock your life-long "skinny gene," then read on … and discover the truth about your innate capacity to reach and stay at a natural weight once and for all.

2

UNCOVERING THE SKINNY GENE DESIGNER'S BLUEPRINT

I have always been in awe of the healing power of the human body. I recall being about seven years old and using my allowance to buy an old medical textbook at a garage sale. Although I couldn't understand many of the words, it lit a spark for a lifelong quest to understand the human body's potential.

In my undergraduate program, I took on a dual major studying cellular molecular biology, emphasis in neuroscience, and psychology, emphasis on biopsychology. When choosing a major, I struggled with the separation between psychology (mind) and cellular biology (body), so I took it upon myself to learn both.

However, the application of my studies was put on hold when, after college, I joined the United States Marine Corps. While on active duty, my interest in the human body's potential shifted heavily toward endurance and physical performance. I was working out five or six days a week, sometimes multiple times a day. Keeping up with the infantry officers (at first), then the pilots in my squadron (some ultra-marathoners) required regular adherence to a strict workout regime.

And while I was busy pushing my body to its physical limits, life delivered a wake-up call. A freak injury resulted in knee reconstruction surgery and my physical performance was suspended. I could not heal my knee faster with exercise, but I could with nutrition. This gave me an opportunity to focus on nutrition. In this period, I became keenly aware of the priority role nutrition plays in healing, so I could get back to that place of high performance.

Fast forward a few years and I was coming back from a deployment in Afghanistan and I was at my leanest weight since high school. However, within a year after returning, I had put on twenty pounds. Other than a few "only get in America" splurges, post-deployment I went back to my normal eating patterns and workout regime. However, suddenly the pounds started slowly creeping up. Then, the more I would focus on either nutrition or exercise, the more I would realize that I was no longer moving the scale. It became clear that something was going on. My body was revealing some sort of imbalance which could not be fixed with a normal, healthy diet and lifestyle. I knew this was *way more* than a "calories-in/ calories-out" situation.

I started dabbling in diets and detox quick-fix programs, but without success. I started becoming frustrated – I'd try on all these diets and protocols to see what would fit, but I had no idea why some worked and some didn't. I didn't know what I was truly solving.

That's when I decided to put my research and cellular molecular biology background to use. I realized that many exposures while on deployment had increased my toxin load and my weight imbalances were reflecting a backup in my detoxification pathways. This finding was validated when my husband and I started having trouble trying to conceive. My hormones were all over the place and I knew that this led back to the liver and the gut. This is where my background in cell molecular and genetics research kicked in and I dug in deep. I started creating a protocol to bring my body back into healing.

After three months on my protocol, I was able to lose weight, get pregnant, and have a successful, healthy pregnancy. I was even able to maintain my weight loss for the last nine years, which included a second pregnancy.

Why did it work? Well, because I decided to roll up my sleeves and figure things out. I realized that no one was going to solve this for me … no one had the answers I had. I needed to start paying attention. It was a combination of paying attention to my body and learning the ways to nourish it best, based on its feedback.

As I was navigating my own journey toward finding my "skinny genes," I had two opportunities to share what I was doing with a family member and a friend to find their own "skinny genes." As a result, one lost fifty pounds and the other lost twenty pounds and was able to overcome fertility challenges. This is when I realized that I wanted to make this passion project into a profession where I could serve others.

In 2015, I started a master's program in Nutrition and Integrative Health that led to completing my graduate degree and becoming a Certified Nutrition Specialist®, the most advanced certification for personalized nutrition practitioners, equipping practitioners to specialize in science-based personalized nutrition therapy.

Working with clients over the last few years, I have discovered that the majority want to lose weight, but above all else, they want to keep it off. Clients crave to have an understanding of their own bodies so they don't keep "yo-yoing." They want results that will last. They want a healthy body and longevity.

After years of watching and observing, the biggest challenge I think clients face in achieving longevity and sustained weight is understanding what their body is telling them. They also don't realize what they are communicating to their body. They are lost and can't seem to achieve their goals because there is literally a disconnect between their mind and their body. How can you possibly get to a destination if you don't know where you are going – and you are driving blindfolded?

My goal is to change this. What I discovered for myself ten years ago wasn't because of my background in science, biology, and neuroscience. My formal education just corroborated what I already knew and you know this too: There is no one like you and there is no one that is going to solve this for you. The answers you are seeking are there all day, every day – within you.

This book is designed to serve as your guide in reminding you of all the ways your body is literally answering all your questions and why you have become so tuned out to its feedback. Through exercises and challenges, exploration, and discovery, I will guide you through designing a diet fit perfectly for you. One that can be continuously shaped and molded as seasons change, life changes, but most of all, a diet that will meet you just where you are at. The more time you give to learning the way your body communicates and how you can communicate back, the quicker you will unlock your skinny genes and keep them on for good!

WHERE GENES MEET DESIGN

What is the framework that makes up your body design?

It really does all start with your genes. However, what is important to understand is that your genes are only a blueprint – akin to a flat, one-dimensional concept. Much like a building, the body is brought to life in full 3D color form by using this blueprint for guidance and direction.

Dr. Ritamarie Loscalzo points out that this blueprint, made up of twenty-three chromosomes and 35,000 genes, is 93 percent similar to rhesus monkeys, 98.5 percent similar to chimpanzees, and 99.5 percent similar to other humans. So how are we so different, so unique? It is your Single Nucleotide Polymorphisms (SNPs) that create the dynamic variety we see expressed. SNPs represent a difference (change or shift) in a single building block (nucleotide) of your DNA.

According to Loscalzo, compared to the 35,000 genes in the human genome, there are 10,000,000 SNPs in the genome. Genes can get turned on or off, but they don't change, whereas the SNPs can get added, deleted, or substituted. SNPs are greatly influenced by the information or instructions from the environment that is created around them. In response to that information, SNPs can impact many metabolic pathways: how nutrients are absorbed or utilized, food molecule sensitivities or intolerances, fat metabolism, detoxification, energy production, blood sugar balances, hormone balance, and neurotransmitter balance. SNPs are the root of shifts that we see in the metabolism. Therefore, the importance is understanding the information or instructions that are guiding the SNPs away from optimal metabolic function – away from our skinny genes.

The information or instructions that are provided to the SNPs to change their expression comes from the environment – both inside and outside the body. This influence has been named epigenetics – a term coined in 1942 to capture the process of influencing the outward expression of genes. This means that we don't have to change our DNA to change the outward expression of our genes, all we need to do is change the environment to signal changes in this expression. This concept was further expanded upon by Deepak Chopra and Rudolph E. Tanzi in their book *Super Genes*, highlighting that only 5 percent of our gene expression is fixed. According to Tanzi, the other 95 percent we get to "mold" through the environment we create, which comes in an abundant form from the nutrients we take in.

In the case of weight gain, there is no greater example of the shift in expression (change in the physical form) than those depicted in identical twin studies. There have been numerous studies that show in adulthood one twin depicting a slender phenotype (physical appearance) and one twin depicting an overweight phenotype. From a genetic perspective, identical twins' genes are exactly the same, but it is the alterations in the SNPs (recall which can be added, deleted, or substituted, based on their environment, inside and outside) that results in the changes we can see in their appearance.

The power of this finding is that it demonstrates that you are not a victim of genetics. There is nothing wrong with your biochemistry and your body is not broken. In fact, your body is still and always will work to its fullest, but can only do so in response to the environment and the information it is being given. Therefore, your power – your ability to create and design your body – comes from your ability to change the environment, the information, the guidance, and direction, to truly transform the body.

WE WERE MADE TO DESIGN

Shaping our biology, physiology, and physical body is our birthright. Our ancestors understood this. As Shanahan describes in *Deep Nutrition*, in ancient traditions nutrition was understood as being a vital tool to shape the best expression of genes. Nutrition was considered "a tool for optimizing human form and function, and for protecting the integrity of family lineage." Somewhere along the way, we misplaced this tool.

Working with a nutritionist trained in nutrigenomics can help you build a plan to address your individual SNPs that are preventing you from completely expressing your skinny genes. However, the level of precision requires genetic testing, and collaboration with a nutrigenomic nutritionist is like jumping into the deep end of the pool. You first want to bring balance to the body through optimizing the way you nourish and becoming aware of the communication between how you are nourishing and your body's response. For many people, learning how to nourish their body, the way it is calling for, and being able to refine based on the response, is enough to unlock the nutrient blocks that are preventing them from reshaping their body and reveal their skinny genes.

NOURISHMENT IS ABOUT BALANCING THE CELL ENVIRONMENT

What really drives this whole process in the body at a basic level is energy, specifically a balance of energy driving the different systems down to the cellular level. Everything from living plants to human beings takes in raw materials and eliminates toxic waste.

Consider these definitions below:

- Nutrient = the capability of anything you consume to be metabolized into a substance that can give electrons or increase energy, this is also often termed an antioxidant, alkaline, Yin

- Waste/ Toxin = directly or indirectly deplete vital biomolecules from having their normal number of electrons, this is also referred to as pro-oxidant, acidic, Yang

This understanding is in contrast to what we normally think of in terms of calories. Weight gain has been culturally accepted to be due to an excess of calories in and not enough calories burned. However, in reality, the metabolic pathways of the body are driven by the duality between nutrients (facilitators) and toxins (which take away or block nutrients). It is about the exchange of energy that is facilitated by the nutrients (optimal performance) or blocked by toxins (impaired function). Therefore, regardless of how little or how hard we "run the engine" (try to burn calories), it is the ability of the system to respond with high performance or impaired function that is the true determinate of weight and balance over time.

We must understand that, as in all systems of the body, we need the duality. For example, in the case of oxygen, reactive oxygen species (ROS) by the definition above would be a toxin and is a byproduct of normal aerobic metabolism of oxygen. According to the *Journal of the American Aging Association*, ROS are important to cellular functioning through signaling and cellular homeostasis when balanced by antioxidants. The issues and damage to DNA occur when the toxins outweigh the antioxidants, preventing normal cell functions. If the toxins continue to outweigh the antioxidants, then it can lead to alterations in the cellular DNA that can be irreversible. It is about proper intake and assimilation balanced and breakdown and disposal (removal) at the right time.

An easy way to think about this is in terms of water. Specifically, hydrogen ions (H+) + hydroxyl group (-OH) = water (H_2O). H_2O is neutral. When there are acids (H+) in balance with bases (-OH) then water is created. When there is too much acid or too much base, then the pH shifts. When the pH shifts, the biochemistry can't function properly and certain systems will demand more nutrients (could be B-vitamins, minerals, oxygen, etc.) to reset the system and bring balance. The body will pull from other areas to support this demand, until either that system gets impacted (no more nutrients to run efficiently) or the extra stores get depleted. In such cases of elevated acid (acidic pH) or depleted nutrients the body will respond with the creation of cravings and in time symptoms to point you toward what the body needs to restore balance. The question is will you receive and understand this message?

This is where your diet design comes in. How are you supporting this balance of nutrients to toxins, acid to alkaline, antioxidant to prooxidant throughout the body, to ensure it has everything it needs to keep the metabolic functions optimal?

INFORM YOUR FORM = DESIGNING YOUR SKINNY GENES

This book is designed to create a template of exploration for you to redefine how you nourish your body, to allow you to start sharpening *your own unique tool* of epigenetic creation to optimize your body's form and function to express your perfect skinny genes. You will be guided through your exploration based on the body's needs (as it is designed). Each step of the way, you will learn how to balance the equation of energy throughout the body by the way you nourish it. You will learn how to pay attention to your body's feedback. I start each of the core chapters with a story about one of my clients, and I'm sure you'll recognize bits of yourself in some or all of them. I hope you find the stories inspirational, because these people were once exactly where you are – frustrated, stuck, and overwhelmed. But they managed to relearn how to listen to and hear their body's innate wisdom for the ultimate weight-loss solution. Throughout the book, you will see references to many studies and current research. All studies and research referenced in the writing of this book are organized by chapter in the *References* section, if you want to take your knowledge deeper and dig into the subjects covered here yourself. In the end, you will have honed your skills in utilizing your nourishment, a.k.a. tailor-made diet, to shape the best expression of your genes and unlock your skinny genes once and for all!

4

STARTING WITH THE "FOOD OF GODS"

Sam opened her mouth for a huge gulp of air. With water rushing down her face, she felt the cool air rush in and fill her lungs. It was a combined feeling of pleasure, relief, and invigoration. Sam finished her laps in the pool and hopped out to start her day. Leaving the gym, she walked outside and when the crisp morning air hit her face, she immediately took in a deep breath and let it out with a long sigh. She noticed the same feeling she experienced in the pool and felt a slight hum in her body. She had a fleeting thought: "If only I could feel like this all day"… the hum quickly faded away as she pictured being stuck behind her desk for the next eight hours.

What was it that caused Sam to feel so good? The answer to this question is found in one single nutrient that the body can't go more than a few minutes without… oxygen.

When it comes to information and guidance to direct our cells, one of the most potent nutrients is oxygen. Here's why it is one of the most critical and often overlooked nutrients in helping us shape our skinny genes.

LINK BETWEEN OXYGEN AND METABOLISM

Oxygen is used for respiration and the release of energy. What many forget is that the cells in your body use the oxygen from your breath to get energy from the food you eat. Oxygen is used to break down sugar (glucose) from our food, then carbon dioxide is produced, and energy is released. This occurs in the mitochondria of the cell, which is often referred to as our energy powerhouse. Think of oxygen as the power source to run the "machine" of the mitochondria to convert food (glucose/ fatty acids) into energy (ATP) through a complex called the electron transport chain. The more power you have in the system, the more energy can be created. So even if you are putting in food sources, if you don't have enough power to convert it to ATP, then there will be a backup in the system.

When there is a backup in the system, this is when modifications of our SNPs occur. In fact, mitochondria are the only cell component that has their own DNA. Research in the scientific journal *Cells* discusses the relationship between mitochondrial DNA (mtDNA) and human DNA in the nucleus of the cell (nDNA). Their research suggests that alterations in mtDNA are primarily caused by oxidative stress, linking oxidative stress to poor energy production. One of the primary causes of increased oxidative stress is more calories than the cell can convert to energy. As we discussed above, the power source to convert food or calories into energy is oxygen.

Without enough oxygen to balance out the carbon dioxide, the mitochondrial system begins to malfunction. It can't process the glucose coming in, so there is a back-up in a step (G6P*) to initiate the glucose to be carried to mitochondria for "energy production."

Getting nerdy with it: Glucose-6-phosphate isomerase is the rate-limiting step that, when shut down, can trigger a decrease in insulin receptors on the cell surface, so no more glucose can enter the cell (can't process).

When energy production is shut down, the cell shifts its receptors to take in less glucose (via decreasing insulin receptors, which shuttle glucose into the cell). This leads to a condition called insulin resistance. Jonathan Temple, a researcher published in the *International Journal of Molecular Sciences*, describes insulin as a "an anabolic hormone that promotes glucose uptake in the liver, skeletal muscle, and adipose tissue." Insulin resistance, triggered by excessive circulating glucose, results in decreased levels of glucose uptake across the body. The body compensates by increasing circulating levels of insulin (increases the signal). As a result, more glucose stays in the bloodstream. In time, this can lead to increased fasting glucose and A1C levels at your next doctor visit.

Additionally, when the mitochondria are compromised or have impaired function due to insulin resistance, then via mtDNA alterations, there is a reduction in the creation of enzymes that activate a pathway, called B-oxidation, by which the body converts fats to energy. What is important to understand as far as fat burning is concerned is that when the body utilizes B-oxidation, it converts one molecule of fat into 127 ATP, compared to glucose metabolism, in which the mitochondria convert one molecule of glucose into four ATP. Therefore, insulin resistance impacts both sources of energy production in the cell. When B-oxidation activity is reduced, then toxic lipid intermediates can build up and create even more oxidative stress and cellular damage.

It is clear to see that the mitochondria have a critical role in energy metabolism and regulating oxidative stress (radioactive oxygen species) production within the cell. Therefore, focusing on nourishing our mitochondria is foundational in designing our skinny genes.

LINK BETWEEN OXYGEN AND DETOXIFICATION AND WEIGHT GAIN

The liver consumes the most oxygen in the body, estimated to be about 20.4 percent of the body's total oxygen consumption, as cited by the University of Hawaii. The liver plays a vital role in balancing the distribution of fatty acids throughout the body. When insulin and glucose increase without uptake (nowhere to go), the liver compensates by increasing lipogenesis (fat stores).

As you can see, these changes in the body are not simply about you putting in too many calories. It is about you needing to turn on the power in the system of the body designed to convert those calories into fuel for the rest of the metabolic processes. Therefore, the signals from the body via elevated blood sugar and weight gain are telling you that you need more power.

Power starts with your breath.

DESIGN CHECK: HOW GOOD ARE YOU AT NOURISHING WITH OXYGEN?

If your body starts giving you any of the following signs, then it is likely signaling that it is time to boost the cell environment and add in information and guidance via oxygen.

- Shallow breathing/ anxious feeling

- Frequent headaches

- Low energy

- Fatigue

- Trouble sitting still

- Racing mind

- Dizziness

What if Sam didn't have to lose that feeling of her body humming, that she experienced when working out, even while sitting at her desk?

We have established that breathing, taking in oxygen, is critical to keeping our physical body going. So, if it is the most important thing to our physical body, then why is it that we spend so little time thinking about getting quality and quantity air? Think about when the body is working hard – do you notice how your body naturally requires more breath? Why is that? It is feedback from the body reminding us that it is an important player in the exchange of energy! We need energy all day, not just when we are working out.

Sam sat down at her desk, opened her computer

and stared at her screen. She saw the clock on the wall to her right and felt her energy and motivation plummet as she considered how many revolutions the big hand would have to make before her day was through. Out of the corner of her eye, she caught sight of her gym bag, hidden discretely beside her desk. She was immediately pulled back to the experience of exhilaration as she took that deep breath during her laps. She closed her eyes and pictured that moment again. This time she took a deep breath in and, recalling being under the water, she held her breath for a moment and then let out a long slow exhale, picturing herself gliding through the water. She then took in a big, deep, cleansing breath and saw herself breaking through the surface at the end of her lap. She opened her eyes. There she was in her office, sitting behind her desk, but that same exhilaration was there. She felt into her body and she could feel a slight hum. Suddenly, the screen in front of her was less off-putting, she felt her energy rise back up and felt ready to take on her day … knowing that any moment she needed to, she could go back to this place through her breath. She considered it her own secret energy treat whenever she needed it. Sam found that the more she practiced her "breath snack," the quicker her day flew by and the more energy she had when she got home, to put toward doing all the things she truly loved – like swimming.

YOUR SKINNY GENES DESIGN KIT

Like Sam, wouldn't you want to have breath snacks if they filled you with energy? Think of this: every time you feel a dip of energy throughout the day and you reach for a snack, as you learned above, it then requires adequate amounts of oxygen to convert the snack into food. What if between your meals and throughout the day, you start with the oxygen first? By giving yourself "breath snacks," you could boost your energy and increase the processing of any food that wasn't yet converted into energy and/ or give your liver much-needed support to facilitate detoxification between meals. I encourage you to start with this simple step-by-step exercise today and observe the changes.

Exercise: Breath Snacks

Step 1: Retraining Your Brain to Breathe for "Breath Snacks"

Many of us are never taught to breathe fully. We take for granted that our body directs us to breathe throughout the day, but how often are we active participants in the process? It is the same as being able to run, but if we want to run further or more efficiently, we need to participate in the trial-and-error process of pushing our body and listening to its feedback to streamline our running stride. Use the steps below to explore with a new way of deep-channel belly breathing, adapted from Dr. Sue Mortor's *Energy Codes*.

- Set an alarm clock for 1 minute. Close your eyes and count the number of breaths in that minute. Write down that number. When we are breathing/ exhaling deeply, we should be taking five or six breaths a minute. The longer the breath, the more oxygen in and the more carbon dioxide released.

- Engage what Dr. Sue calls "Mula Bandha/ Root Lock." To do this, you clench at the base of your pelvis (like you are doing a Kegel exercise or holding from having to go the bathroom). Ensure your shoulder blades are pulled together toward the spine, down, and away from the shoulders. Pull in at the front of the chest like you are hugging your heart.

- Pull your breath down from a point about 2 inches up to 2 feet above your head. Pull that breath down into an opening in the top of the crown of your head.

- Picture a funnel that expands right at the belly area (3 inches below your navel) and runs down from your body into the ground. As you breathe in from the crown, see that energy pouring into the center of the funnel.

• On the exhale, let the energy move from your navel right down into the center of the earth. Just relax and let it drop.

• On the next inhale, breathe that energy back up from the earth through the funnel, running through the heart, up through the throat, through the center of the brain, and out the top of the head to the point 2 inches to 2 feet above the crown. Picture the energy rising like a whale shooting water out of its blowhole.

• Repeat. This time on the inhale, pull the energy from the crown down and into the belly and let the belly relax and expand out fully. On the exhale, let the belly compress deep into the spine and shoot the energy down and into the earth.

• On the next inhale, draw the energy up from the earth into the belly and let the belly relax and expand out fully. On the exhale, let the belly compress deep into the spine and shoot the energy back up through the sternum, through the heart, throat, third eye, and out through the crown of the head.

- Once again, set a timer for 1 minute. Close your eyes and count the number of breaths in that minute. Write down that number.

This practice cultivates more high-frequency energy in the body. When we have a lot of emotions constricting the body, this becomes harder to do and it may be challenging to get the belly to fully expand.

If you find it challenging at first, place your hands on your belly so you can feel the resistance, the expansion, and the compression until it becomes easier.

It also can be hard to continue to contract Mula Bandha while allowing the belly to expand. It is this push/ pull that is a powerful part of the practice. So, it may require frequent practice in order to get the full benefit of the breath.

Step 2: Designing Your "Breath Snack"

Practice: Commit to five days in a row to conduct this exercise. Then set windows during the day, possibly when you normally reach for a snack – when you are going to indulge yourself in a rich snack of oxygen. Throughout the week, you can expand these windows to include a "breath snack" first thing in the morning or right before bed. During this time, you are going to set an alarm for the following durations. Day 1 = one minute, Day 2 = two minutes, Day 3 = three minutes, Day 4 = four minutes, Day 5 = five minutes.

Aim to get one of these "snacks" in fresh air, outdoors, in nature (if you can). Throughout the week, this time will be dedicated to deep, central-channel belly breathing. Follow these steps:

1. Rate your energy level on a scale from 1-10.

2. Conduct the deep breathing practice from your baseline exercise above. As you are breathing, visualize the oxygen going to each cell of the body. When you are done, note your energy level.

3. Note your energy level on a scale from 1-10.

4. At the end of the week, take a few minutes to journal on the following questions:

- How has my energy been affected by dedicated breath snacks? Did I feel the need to eat in the 1-2 hours after my "breath snack?"

- How has my mood been affected?

- How did I handle stress?

- How were my food choices impacted?

- How was my digestion impacted?

5. Review the results of your self-assessment.

The key is observing how you feel: hunger/ cravings, energy, sleep, mood, stress…. Make a commitment to yourself that you will do this practice for seven days straight until it becomes a daily part of your nourishing design to unlock your skinny genes.

5

GETTING INTO THE FLOW

Alex started talking with her colleagues at work about strange smells coming from the sewers outside the office. One of her colleagues joked about it smelling no worse than the water in the river running through their city. He went on to say that it was filled with toxins and can't imagine what was making its way into the tap water. Alex felt herself getting defensive and immediately responded, "There is nothing wrong with tap water, I drink it every day". Later during a break between meetings, she stared at herself in the bathroom mirror and couldn't shake the conversation from earlier ... was there something she was missing in her water? Could it be toxic? Was this a contributing factor to why she couldn't meet her weight loss goals? She thought back to when she was her leanest five years earlier ... before she moved to the city. Her roommate at the time was obsessed with purified water and she took advantage because something about it just tasted better, but Alex hadn't really given it much thought. She decided maybe she would do something different and stop drinking tap water. Could it really make a difference?

Water is the second source of nutrition that the body cannot live without.

You may think, "I know I need water." You may also recall learning that your body is made up of about 60-70 percent (depending on the source) of water. But take a second and ask yourself: what does that water do? Is there a single organ in your body that can function without it? When we were growing in our mother's womb, was there any part of us that wasn't bathed in fluid, predominantly made up of water?

Television shows like *Alone, Survival Man*, and *You vs. Wild* remind us that the body cannot go for more than a few days (estimated three-five days) without water. Our organs literally start shutting down when we are deprived of this vital nutrient. Therefore, water is the second most important nutrient to nourish with, based on the rapid feedback from the body when we do not have it. If the body reveals in extreme conditions that it is critical to thrive, then how important is it in our daily lives? If you are like most people, water isn't high on your priority list. Even if you are conscious about water sources, do you focus on meeting your daily requirement of water? Do you know what your daily requirement is?

Let's explore ...

DAILY CLEANING

In the last three years, a study revealed that the largest organ in the body isn't actually the skin but rather the interstitium, comprised of fluid-filled spaces "within and between tissues all over your body." It is in this system that a majority of the "cleaning" occurs. In interstitial space, the cells are bathed and nourished with oxygen, glucose, amino acids, and other nutrients needed by tissue cells. According to Robert O. Young, author of the book *The pH Miracle*, it is also within this space that the acidic waste products of the cell and the blood are diffused and drain toward the lymph nodes – an immune system gland that filters clear fluids. From the lymph nodes, the waste can be eliminated via the skin, urinary tract, or bowels. However, before it is eliminated, Young describes that the metabolic and gastrointestinal waste acids are stored in this space until they can be buffered and eliminated.

The study found that the interstitium collapses due to dehydration, indicating the critical role that hydration plays in this new-found organ. Furthermore, the body reveals the connection between water and our body's ability to "clean house."

LINK TO WEIGHT GAIN

What happens if there is a buildup in toxins? Water is created as a byproduct of energy (ATP) production in the cell. Therefore, having enough oxygen, as mentioned in Chapter 4, leads to increased water in the interstitium. We contribute to this through the amount of water we consume on a daily basis. If either source of water (generated by the cell or consumed) is not at a balanced level to meet the demands of "cleaning house," then we create an electrolyte imbalance.

The water in the extracellular (interstitial) fluid is tightly regulated by buffers and maintained around a 7.4 pH, but there are limited buffers available. When there is mitochondrial dysfunction (as referenced in Chapter 4), this leads to an overproduction of hydrogen ions (not enough oxygen to bind to in order to form water molecules). Think of "too much" in terms of excess toxins or too much "trash build-up" without a trashman to collect it and remove it to the landfill.

What happens if there is too much trash on the street corner? If you have no place for it to go? What would you do if your trash was piling up? Would you consume fewer items that create trash? This is how the body responds. When the pH of the interstitial fluids is low (in a state of acidosis) there is a reduced binding affinity of insulin to its receptor on the cell surface. Insulin is what brings glucose into the cell. However, if the cell can't eliminate toxins, then it is not in a state to take in nutrients to create energy (it needs to clean house first). This means that the cell shifts its receptors through the feedback mechanisms and intra-cell communication to prevent taking in more glucose (that when consumed will lead to more "trash"). This results in increased insulin in the bloodstream and decreased glucose uptake, referred to as insulin resistance, which is associated with weight gain.

It is also important to understand how the body eliminates the toxins or "trash" from the cell. The interstitial fluid drains into the lymph. Recall that in Chapter 3, we discussed the two critical functions of all cells: to consume and assimilate nutrients and break down and eliminate waste. Alongside waste from the consumption of food, the by-products (waste) of our 37.2 trillion cells need to be eliminated. The lymphatic system is the sewage system of the body.

This waste build-up can create a vicious cycle of toxin load (too much waste) and not enough elimination. A study in 2013 observed that weight gain, which can come from insulin resistance, resulted in significantly impaired lymphatic fluid transport and lymph node uptake. The same study revealed that such impaired lymphatic function altered the expression (genetic polymorphisms) of lymphatic markers that are present in the lymphatics of lean counterparts. Therefore, built-up metabolic "acidic" waste that is not balanced with alkalinity can lead to a decreased ability to eliminate the waste that is going to increase body storage of toxins, primarily in adipose (fat) tissue.

If you want to balance your blood sugar levels and keep your system "clean" and lean, consuming your daily optimal water dose is a key factor in unlocking your skinny genes!

CLEANING WITH PURE WATER

Now imagine that you are cleaning your house. What would happen if you started with a bucket that had remnants of mud and dirt from a previous cleaning? You fill it with water, but the water comes out brown and has even more mud, dirt, and debris in it. How effective would your cleaning efforts be? Furthermore, as you continue to clean, you keep using the same water and don't refill, only concentrating the mud and dirt until it is thick and sludge-like. Would you be able to keep your house clean this way? The easier answer is no. This is no different from our system internally. All day long, the body smartly prioritizes a time for "house cleaning."

If your water is filled with toxins, chemicals, and pollutants, then it is more toxic to start with. This is akin to cleaning the house with dirty water. In recent years, tap water has been identified as a significant source of toxicity. Various reports and forensic analysis have identified the following contaminants in water: heavy metals (e.g., lead, arsenic, cadmium), fluoride, chlorine, solvents (e.g., trihalomethanes class includes chloroform), more than 600 disinfectant by-products, bacteria, parasites, phytoestrogens and xenobiotics, pesticides (e.g., glyphosate), insecticides (e.g., dichloro-diphenyl-trichloroethane [DDT]), fungicides (e.g. hexachlorobenzene [HCB)]), and herbicides (e.g., dimethyl-tetrachloro-terephthalate [DCPA]). All these contaminants in water put a burden on the body's nutrient-toxin load. In the article "Why There Is No Such Thing as Safe Tap Water," Sayer Ji highlights the issues with contaminants such as DDT, HCB, and DCPA – they are biopersistent chemicals, meaning that they are resistant to breakdown through metabolic processes and persist in adipose (fat) tissue, where the body places them if they can't properly and safely be eliminated. The synergistic effect of these contaminants creates a significant burden on the body, increasing the toxin load, and creating an environment for the cell that can lead to genetic polymorphisms that shift the body away from optimal biologic function.

In order to properly "clean house" daily to support reducing your toxin load, you need clean water. The way to get this and to remove all of the contaminants mentioned above is through reverse osmosis water or distilled water placed through a solid carbon/ fluoride filter. In terms of the energy balance, reverse osmosis has the greatest alkalinity ratio of the three water sources. There are many options for people to purchase reverse osmosis water or distilled water types and then place them through a pitcher or countertop solid carbon/ fluoride filter. Spring water likely contains many of these contaminants and therefore requires additional filtration.

Additionally, to avoid added xenobiotics, it is important if purchasing water to get it in BPA-free bottles or preferably glass. The same is true for your own personal water storage for daily drinking bottles.

Although it is a couple of extra steps to clean your water and an investment of money for water or filtration systems, it may be helpful to think of water as a nutrient (a part of your nourishment/ food budget). Also, think of all the effort, energy, and time you are saving if you start with clean water every day, to not have constantly offset this toxin burden through other nutrients on a daily basis.

WATER BEYOND BIOCHEMISTRY – THOUGHTS/ EMOTION IMPRINTING

It is well understood in ancient traditional healing modalities (i.e., traditional Chinese medicine) that there are energy paths that transverse throughout the body. Furthermore, after thousands of years of accumulating data, it was found that certain thought patterns and emotions were aligned to imbalances in specific organ systems. The connection between these emotions and the physical shift in the biochemistry affecting organ systems is not well understood.

However, advances in technology have led to some profound discoveries that suggest water has a critical role in mediating this interaction between the mind and the body. Specifically, the work of a Japanese scientist by the name Dr. Masaru Emoto used water crystal photography to capture the structure of water after exposure to music, words (written or verbal), and mind's intent. His work revealed that water "exhibits a living quality" through its ability to transform between a beautiful, well-balanced hexagonal crystal and a badly formed, unbalanced crystal based on the type of energy imprinted in it.

Emoto's photographic capture of the changes in water structure provides visual evidence of the previous experiments of Smith and Tiller, who theorized that water had a memory for frequencies and changed its structure according to the frequency imprinted. While Smith and Tiller focused primarily on intention, Emoto was the first to expand the imprinting exploration to include music and words. The ability to imprint water with music and words that can change it into a more coherent state with crystalline structure implies the ability to optimize water utility in the body. As discussed above, many of our water sources are contaminated and this causes the water molecules to move in a chaotic and irregular manner, losing their coherent quality. It is this coherence that enables the constant communication of molecules through our cellular structure and our DNA.

This information exchange is facilitated by water molecules. Therefore, if we want to optimize the communication, information exchange, and genetic expression of our cells toward our skinny genes, then it is essential that we ensure the water in our body is in its most coherent state.

DESIGN CHECK: HOW GOOD ARE YOU AT NOURISHING WITH WATER?

We have established that having more alkaline and cleaner, clearer (not bogged down by toxins, chemicals, synthetic byproducts, etc.) water in our interstitium is required to keep the powerhouse of our body (mitochondria) at highest capacity in order to keep our system in balance. However, it appears that what is equally important is how we nourish ourselves with thoughts, emotions, words, and songs. The research is showing that these things don't go out into the air and disappear, but rather they bathe our cells. Since our body gives us feedback on our current state, if we don't make an effort to put in fresh, clear water each day, then we likely will continue patterns of thoughts, emotions, and words that don't serve us and in time create dis-ease in the body.

Here is an easy body feedback tool to determine if you are hydrating properly.

Dehydration test:

- Stand with your hands at your sides.

- Palpitate the veins in your right hand with your left hand.

- Raise your hands to heart level to see if the veins are still visible or not.

- If not, you are dehydrated.

You can do this at any time you wish to determine your hydration status. Test it out at different points in the day. You can also use it to see how sensitive your body is to challenges such as exercise. Tuning into your body to learn when you need to fill up the water bottle can be a powerful tool in your skinny gene design.

What if Alex saw water as a powerful tool to shift her energy and her thoughts?

Alex took a big sip of refreshing cold water. She reflected on how clean water really does taste better. It had been three weeks since she started using filtered water and she had already noticed more energy. She was surprised to see she had lost 2.5 pounds in the first week, without changing anything but her water source. Additionally, she had started exploring with water imprinting. She knew her confidence was an area that she needed work, so before her last big meeting in which she had to propose a new project idea, Alex had placed yellow paper written with the affirmation: "I open my mind to possibility" on the inside of her water bottle holder. She didn't really expect anything, but was open to seeing what happened. To her surprise, not only did Alex feel at ease and confident briefing, but everyone was receptive to her idea. Most importantly, her office chief – whom Alex was aware had a recent track record for shutting down innovation and being reluctant to spend money – said he would be willing to fund the project. Alex was thrilled. It felt so easy. It felt like her body and mind were humming at a level she hadn't experienced for years.

YOUR SKINNY GENES DESIGN KIT

Like Alex, wouldn't you want to have optimal water cleaning out your system daily and boosting your mind-body state as a regular part of your skinny gene design?

Exercise: Powering Up Your Water – Seven Days

Step 1: Calculate Your Daily Hydration Requirement

- Total requirement: Divide your current body weight by two, the resulting number is the number of ounces you need to consume in total.

- Factoring for losses: Even if you have met your daily requirement, use the guidelines below to replace lost water through other beverages, exercise, etc.

  ○ 1:1 for caffeine. For each one ounce of caffeine consumed, replace it with an equal ounce of water.

  ○ 1:4 for soda. For each one ounce of soda consumed, replace it with four ounces of water.

  ○ 1:16 for exercise. Calculate your sweat rate during exercise: Take your weight immediately before and after working out. For every pound of weight loss, replace with sixteen ounces of water.

Step 2: Daily Power-Up: Exploring with Water Imprinting

Using a glass water bottle, tape a piece of paper with the list of colors and affirmations listed by day, below. Affirmations are attributed to Dr. Sue Morter's book *Energy Codes*.

- Day 1
Color: Red
Affirmation: "This is my gig," "I belong," "I bring what I choose to experience."
Music Key: C

- Day 2
Color: Orange
Affirmation: "I sense and feel my way through life," "I need nothing from you, I am simply here to share," "I follow my gut"
Music Key: D

- Day 3
Color: Yellow
Affirmation: "I allow my own way, and allow you yours," "I open my mind to possibility"
Music Key: E

- Day 4
Color: Green
Affirmation: "There is more than enough for all,"

"There is only one of us here – we are one,"
"Everything is a reflection of the divine, and it is in my favor"
Music Key: F

- Day 5
Color: Blue
Affirmation: "I hear and speak the truth with love and compassion," "I manifest myself here fully," "My life is a reflection of my inner work"
Music Key: G

- Day 6
Color: Indigo
Affirmation: "I am the one behind the eyes"
Music Key: A, specifically 432 hertz

- Day 7
Color: Violet/White
Affirmation: "I am a divine being," "I am that," "Life is a reflection of all that I am"
Music Key: B

Adding frequency. Building on Emoto's research, you can try exploring with frequency as well. Conduct a Google search or use Song Key Finder (https://www.songkeyfinder.com) to find music in your favorite genre in the key aligned to the day and put it on at some point during the day. Put your water near the music source, if possible.

At the end of each day, do a five-minute journal exercise connecting with your experiences from your daily practice and reflect on how your body felt throughout the day. Was there anything that shifted? Was there anything that was challenging? The key is observing how you feel: hunger/ cravings, energy, sleep, mood, stress.

Based on this exploration, how do you view water's role in your energy balance? Imagine how you will use water as a daily part of your nourishing design to unlock your skinny genes.

6

PRIMING FOR THE BURN

Megan had always struggled to keep her weight in balance. If she gained even five pounds, she felt like everything in the world was wrong. She found that if she didn't keep to a strict diet and lost a few days of running, then the excess pounds would creep back up. Finally, one day she pushed too hard and hurt her knee. She was told by the doctor she had to stay off it for at least ten days. This filled her with dread. Megan knew that she would gain at least five pounds in that time if she wasn't working out and all her clothes would fit wrong. She went home with a racing mind. She pulled in the driveway and took note of the access to a trail system at the end of her neighborhood. The doctor said she couldn't run but she could definitely still walk. So, she took a walk to clear her head. She got to an opening on the trail and sat on a rock for a minute, giving her knee a break and letting the sun soak into her skin. She got lost in the moment and the world and the worries of her weight – something she could never seem to shake – seemed to fall away. She closed her eyes and suddenly, as if for the first time on the walk, heard the birds singing. Before she knew it, fifteen minutes had passed. She got up to head home and all the while enjoyed the sun, trees, and birds. She went about her day and decided to not let her fears get the best of her. She made a commitment to spend a few minutes every day walking on the trail and sitting in the sun.

When it comes to information and guidance to direct our cells, another often overlooked nutrient is vitamin D. Research has revealed that vitamin D is a fat-soluble vitamin that functions as a prohormone (hormone precursor) and regulates the expression of an estimated 900-1000 genes.

Vitamin D's role in genetic regulation is exemplified in pregnancy – a critical period of optimal development. Studies have demonstrated that vitamin D supplementation during pregnancy alters transcriptome (range of messenger RNA) and epigenetic alterations through DNA methylation in genes that regulate metabolism, immunity, inflammation, cellular health (proliferation vs. death), nerve impulses, brain tissue development and differentiation, and sensory organ development. These findings demonstrate the foundational role that vitamin D plays in our initial development and body design as fetuses in utero. Additionally, Bruce Hollis, a researcher who has led extensive research on the implications of vitamin D status during pregnancy, identified that during pregnancy the vitamin D requirement for women is increased in the body due to an added pathway dedicated to vitamin D and its metabolites for genetic signaling. The placenta has been found to express high levels of a vitamin D-activating enzyme and vitamin D receptor (VDR). These modifiable effects of tissues during pregnancy to increase expression of vitamin D and its metabolites depict the body's prioritization of this nutrient in optimal health.

LINK TO WEIGHT GAIN

Vitamin D and the Microbiome

The vitamin D receptor (VDR) is a protein that controls the rate of transcription (making copies) of genetic information from DNA to messenger RNA. In complex with hormonally active vitamin D (1,25-OH), VDR regulates the expression of more than 900 genes involved in a wide array of physiological functions. VDR mediates most of the known functions of vitamin D (D_3), which include immune function, inflammation, insulin-like growth factor signaling, sex hormone signaling, xenobiotic detoxification, mineral and bone metabolism, cardiovascular function, regulation of cell proliferation and differentiation, calcium metabolism, and calcium absorption. Low VDR signaling transduction is linked to shifts away from a healthy balance of commensal bacteria in the gastrointestinal tract. Bacterial infections (e.g., food poisoning), parasites, viruses, and yeast/fungus overgrowths due to shifts in gastrointestinal pH can lead to impaired intestinal terrain and can create an environment favorable for microbes that require immunosuppression to achieve survival. They do this by subverting (downregulating) one of the body's most prolific nuclear receptors, the vitamin D receptor (VDR), responsible for expression of several families of key endogenous (made naturally within the body) antimicrobials. Since VDR also assists in the maintenance of gastrointestinal integrity, helping prevent intestinal hyperpermeability, subverting VDR, and decreased signaling are associated with intestinal permeability and autoimmune conditions.

A researcher and professor at the University of Illinois, Jun Sun, lays claim to the key achievement of characterizing the vitamin D receptor and its regulation of the microbiome. In her work, Jun showed that the human VDR gene is the key factor in shaping the gut microbiome. As you will learn in Chapter 8, the microbiome is the "heavy lifter" when it comes to maintaining a lean figure and expressing our skinny genes. Jun found that the absence, and therefore inactivation, of VDR receptors creates dysbiosis (imbalanced microbiome), contributing to inflammation in the intestinal epithelium (the lining of the gut) and increased metabolic endotoxemia (accumulation of lipopolysaccharides [LPS]). This dysbiosis in combination with a high-fat, high-processed food, high-sugar diet led to insulin resistance and weight gain.

The significant impact of vitamin D deficiency on the integrity of the intestinal tract is further substantiated by Ann Louise Gittleman, author of *Radical Metabolism*. She highlights a study on Crohn's patients, who have extremely injured and hyperpermeable GI tracts. They were provided 2,000 milligrams of D_3 supplementation per day. The supplementation resulted in reduced intestinal hyperpermeability. As you will see in future discussions in Chapters 8 and 9, intestinal permeability and inflammation are an underlying root for weight gain.

Vitamin D and the Liver

Additionally, the shift in microbiome composition due to vitamin D deficiency reduces the abundance of the flora that produces short-chain fatty acids (SCFA), vitamin K, vitamin B12, biotin, folate, and thiamine (vitamin B1). These vitamins are cofactors in detoxification pathways and without ample amounts for use by the liver, it can lead to critical enzymes becoming downregulated (decreased). For example, vitamin D_3 is activated in the liver by cytochrome P450 (CYP) enzymes, when these enzymes become downregulated (decreased) due to an increased toxin load, then the level of active vitamin D (1,25-OH) is decreased.

However, active vitamin D (1,25-OH) is essential for maintaining optimal enzymatic function of detoxification pathways (see Chapter 9 for a greater discussion on these pathways). One study found that within six hours of a single dose of activated vitamin D (1,25-OH), antioxidant genes increased expression to support greater detoxification. These findings demonstrate that vitamin D plays a critical role in supporting detoxification and protecting against environmental toxins that can contribute to weight gain.

Another significant demonstration of vitamin D's ability to protect from toxins and alleviate the burden on the liver is found in its ability to protect from endotoxins, i.e., lipopolysaccharides (LPS), that make their way into the bloodstream from gram-negative bacteria in the gastrointestinal tract. A research study found the vitamin D and its metabolites suppressed LPS-induced cytokine and chemokine expression. LPS has been linked to weight gain associated with metabolic syndrome. The role of vitamin D to suppress this endotoxin demonstrates just one of the mechanisms by which it supports the body in reducing triggers of weight gain.

Your role is to ensure you are building vitamin D into your daily design to support a healthy microbiome and intestinal strength to keep VDR receptors supporting detoxification, mineral metabolism (see Chapter 7 for more details), and balanced cellular proliferation to optimize your skinny genes.

DESIGN CHECK: ARE YOU GETTING ENOUGH VITAMIN D?

Throughout most of human history, vitamin D was obtained almost exclusively from cutaneous biosynthesis, since typical diets contained little or no vitamin D. Moreover, people who are exposed to adequate amounts of sunlight do not require a dietary source of the vitamin. During the months of spring through fall, March through October, it is recommended to sit in the sun for ten to fifteen minutes, two or three times a week, between the hours of 10 a.m. and 3 p.m., with your arms and legs or hands, arms, and face exposed to the sunlight in order to create adequate vitamin D production. The challenge arises when either there is not an opportunity to get sunlight exposure during these windows, or in winter months when there are low UV rays.

A recent study in *Nutrients*, a peer-reviewed journal of human nutrition, estimated that one billion people worldwide are vitamin D-deficient. A rise in work performed indoors, body and facial coverings, and exclusively using sunblock when outside has contributed to this deficiency. Additionally, optimal UV rays to support adequate production are only viable during the spring and summer months, which leads to people requiring added dietary and/ or supplementation sources to maintain optimal levels in fall and winter months.

This is where nutrition can come in. Vitamin D_3 occurs naturally in fish and in small amounts in a few other foods (e.g., fish (salmon and sardines), fatty fish liver oils (i.e., cod liver), grass-fed animal fats (lard, tallow, etc.), pasture and free-range eggs, grass-fed butter, grass-fed beef liver, and grass-fed cheese.), and is synthesized in the skin after exposure to sunlight or other sources of ultraviolet light. Vitamin D_2 (also called ergocalciferol) is produced by irradiation of ergosterol, a sterol present in fungi. Vitamin D_2 is not normally present in the human body and is found only in trace amounts in plants, such as mushrooms. However, Vitamin D_2 is still capable of inducing vitamin D activity in humans and has been used for decades to prevent and treat vitamin D deficiency.

The recommended daily allowance (RDA) for adult vitamin D supplementation is 800 IU. However, most doctors, researchers, and nutritionists will say that adults of normal weight should get a minimum of 2000 IU per day, on average. The challenge with vitamin D supplementation is that it is a fat-soluble vitamin. This means that if you have any issues with fat metabolism (see Chapters 8 and 9) then you may not absorb the amount you are supplementing. According to Dr. Holick, director of the Vitamin D, Skin, and Bone Research Laboratory at Boston University School of Medicine and author of *The Vitamin D Solution*, since vitamin D is fat-soluble, it gets diluted in body fat, requiring increased food and supplementation to satisfy the body's daily requirement.

Based on the biological and epigenetic importance of this nutrient to help regulate many metabolic functions, it is important to know your status. The best way to determine your need for vitamin D is to work with your doctor or nutritionist to have your vitamin D levels tested (both 25-OH and 1,25-dihydroxycholecalciferol) in order to determine the precise levels for you.

Did Megan just need a little bit of sunshine in her day?

Megan had kept to her commitment and for seven out of the ten days, missing three days due to rain, she walked on the trail mid-day and took time to be in the sunshine. She also had made a commitment to herself that she would not get on the scale for the entire ten-day recovery period until her doctor said she was cleared to start running again. Informed by previous experience, she was rightfully afraid that the number on the scale would make her spiral. So when she had finally got the green light to get back to working out, she pulled out the scale and with a little reluctance stepped on to see what added weight she now had to figure out how to burn. To her utter astonishment, she didn't end up gaining 5 pounds but found that she actually had lost 1.5 pounds. She was so taken aback. Could it be that she had it all wrong this whole time? Could working out really not be the total solution from her meeting and sustaining her ideal weight? Whether it was the fresh air or sunshine, Megan didn't care – all she knew was that she had found a new way to help herself stay slender and it was a habit that she intended to keep.

YOUR SKINNY GENES DESIGN KIT

Like Megan, are you missing optimal doses of vitamin D to prime your metabolism to burn? Try exploring how adding in a daily dose will influence your skinny gene design.

Exercise: Daily Dose of Sunshine

- Spring and summer (April-September): Leverage the sun. Expose your face, neck, and arms to the sun for fifteen to twenty minutes a day in the morning (between 10-12), preferably in nature.

- Fall and winter (October - March): Especially in northern latitudes, little light at the right wavelength comes through the atmosphere in the fall and winter to support vitamin D absorption via sunlight. Therefore, during these months, it is best to eat vitamin D-rich foods and to supplement. It is also optimal to have your vitamin D levels tested (both 25-OH and 1,25-dihydroxycholecalciferol) and work with your doctor or nutritionist to determine the right levels for you.

• All four seasons: Explore by adding in one food that is rich in vitamin D.

○ Add in D_3 from animal sources: fish (salmon and sardines), fatty fish liver oils (i.e., cod liver), grass-fed (GF) animal fats (lard, tallow, etc.), eggs, GF butter, GF beef liver, and GF cheese.

○ Add in D_2 from mushrooms: maitake, morel, chanterelle, oyster, and shiitake all contain natural levels of vitamin D_2.

○ Add in daily greens and other plants to get vital minerals required to utilize vitamin D to optimize metabolism, cell growth, and repair (see Chapter 7 for more details).

The key is observing how you feel: hunger/ cravings, energy, sleep, mood, stress. If you notice an improvement, explore with regular daily sunbreak, including other foods rich in vitamin D, or work with your nutritionist to determine optimal supplementation so that this critical nutrient becomes a daily part of your nourishing design to unlock your skinny genes.

SPARKING THE FIRE

Liz stared at the page of her book and realized she had read the same page three times without being able to process any of it. Twenty minutes had passed and she still wasn't sleepy. For a second, she considered getting up to get a piece of chocolate, just a tiny piece; her mouth watered from the very thought. "Nah, all the white-flour pastry I grabbed at work today ... I really shouldn't." She felt exhausted, but it was like her body couldn't calm down. Her legs felt achy and she found herself continuing to move them to get comfortable. Holding the book in front of her, she noticed another fingernail had chipped. She could never seem to keep her nails looking good. Her mind was filled with all the money she had spent over the years on products to keep her nails strong, manicures and polish were only ever temporary cover-ups for they inevitably chipped again.

"I guess it is not in the cards for me to have beautiful nails ... just another way my body is broken," thought Liz as she shut out the light and gave up the effort of trying to lull herself to sleep by reading. As Liz lay there trying to sleep, she couldn't help wondering if her body was really broken or was there something more going on? Could it be that her trouble falling asleep, achy body, and chipping fingernails were all related? If so, what was it telling her ... was this somehow linked to her struggles with losing weight?

In balancing the cellular environment to optimize our SNPs, minerals play an important role in the utilization of nutrients and detoxification of toxins. They are critical cofactors (or catalysts) in hundreds of metabolic functions. When there are not enough minerals to support the pathway, then it gets back up. Although we need a wide variety of minerals daily through our diet, there are two minerals that we will focus on in detail, due to the vitally important roles in metabolism, but also since the majority of the population has been found to be deficient in these nutrients: magnesium and boron.

SOIL ESSENTIALS: LEARNING FROM NATURE

Just as plants need minerals from soil to utilize sunlight to optimize plant growth and repair, you need minerals from plants to utilize sunlight (vitamin D) to optimize metabolism, cell growth, and repair. The two minerals that help you absorb vitamin D are magnesium and boron. However, in the last few decades, due to modern farming techniques, both magnesium and boron have become some of the most deficient minerals in our depleted soils.

Dr. Alan Gaby, nutrition guru and author of *Nutritional Medicine*, suggests the use of ammonia as fertilizer unbinds magnesium in the soil. Studies have revealed a 21-35 percent decline in the magnesium content of vegetables and fruit. This means that even if you consume a diet rich in vegetables and fruit (intended to consume these minerals), it is assessed by some reports that you are getting only 40-50 percent of what previous generations used to get and what your body's metabolism is designed to relying on.

As you will see below, the body has a need for magnesium and boron. Lara Pizzorno, a researcher who wrote in the *Integrative Medicine Clinical Journal* about boron, highlights how these minerals play vitally important roles in metabolism and in supporting the body's natural production of antioxidants to support daily balancing of the nutrient-to-toxin load. Now consider the increase in pollutants in our environment and the decrease in our body's consumption of magnesium and boron. Over the last ten to fifteen years, as a population, have people gotten leaner or heavier? Could these essential minerals missing in our diet be a contributing factor?

SKINNY ON MAGNESIUM

Some of the first receptors in the intestine, where you begin absorbing nutrients, are aligned with critical minerals, such as magnesium. Magnesium is a cofactor for 700 life-sustaining reactions happening in your body, and is responsible for more than 350 functions in your body. In the body, magnesium receptors are found across the small intestine (where the majority of nutrient absorption takes place) in both the upper and lower duodenum and the ileum. No other mineral has this many sites of absorption. This signals that it is an important mineral in the body. If absorption of this mineral is a priority for the body, then it should be a priority for your skinny gene design.

Magnesium is one of the most overlooked and yet most vital minerals in the body and is likely to be one of the most frequently encountered nutritional problems found in clinical practice, according to Dr. Alan Gaby. This is due to magnesium's essential functions: it serves as a cofactor for more than 300 different enzymes, Dr. Gaby says, and is essential to ATP synthesis or energy generating. As you may recall from the discussion in Chapter 3, ATP is universally seen as the energy exchange factor that provides the cellular energy for building, breaking down, movement, and transport. Dr. Dennis Goodman, author of the book *Magnificent Magnesium*, also points out that magnesium binding sites have been found on "3,751 human proteins that are essential for building, repairing, and maintaining your body's cells."

It is assessed that approximately 80 percent of American adults are still magnesium deficient. In addition to depleted soils and low nutrient density in our foods, we regularly consume foods that prevent optimal absorption due to refinement and processing of foods. For example, Gaby finds "85 percent of magnesium is lost in the refining of whole wheat flour to white flour." Adding insult to an already nutrient-depleted situation, the consumption of beverages such as soda, that contain phosphoric acid, high-fructose corn syrup, and other additives, can also deplete magnesium. Alcohol consumption directly interferes with the body's ability to absorb magnesium through the kidneys. Other dietary factors that can deplete magnesium are coffee and caffeinated beverage consumption, which lead to nutrient flushing; lack of healthy fats (including animal fats), which reduces mineral absorption and transport; and high consumption of acid-forming foods (i.e., meat, sugars, simple carbohydrates), which raises pH and requires magnesium to neutralize excess acid, according to Goodman.

LINK TO WEIGHT GAIN

Magnesium is a rate-limiting step in glycolysis, the process by which cells make energy (ATP). Due to its role in the creation of energy, magnesium deficiency has been linked with insulin resistance and risk for type 2 diabetes mellitus and metabolic disorders. This correlation is likely due to magnesium's ability to mitigate stress and toxins to bring balance to metabolic functions, as outlined below.

It is common knowledge that stress can lead to health imbalances that include weight gain, but what many people don't realize is that one of the key linkages between the two is magnesium. Magnesium also works directly at the Hypothalamus-Pituitary-Adrenal (HPA) axis to suppress cortisol and adrenal output, two of the body's "stress hormones." It works by reducing the release of the adrenal stimulating hormone ACTH (adrenocorticotropic hormone) from the pituitary. Plus, it has the unique ability to cross the blood-brain barrier, serving as a gatekeeper to stress responders and blocking the entrance of stress hormones into the brain. As we will discuss further in Chapters 9 and 10, the hypothalamus controls the body's appetite and energy homeostasis. This means that based on feedback from chemical messengers in the body, it will influence the body's basal metabolic rate, appetite, and body temperature. It is the hypothalamus that controls if your body is shifted toward a state of starvation (in which it needs to store energy) or abundance (in which energy is readily available).

According to Goodman, stress depletes the body of magnesium due to magnesium's role in regulating hormone production. Goodman goes on to explain that when we are in a state of chronic stress with elevated levels of circulating stress hormones, magnesium is quickly expended to act as a buffer and to bring hormone levels into balance. Therefore, the more stress hormones, the more magnesium is expended, leading to chronic magnesium deficiency, "cellular energy loss," and in time, disease.

Magnesium also plays a role in a balanced microbiome and Dr. Kristin Schnurr makes this connection in her article "'Magnesium and Hormonal Balance" in which she shows clinical correlations between magnesium deficiency and altered gut microbiota. Specifically, she found a magnesium-deficient diet for six weeks significantly increased immobility time (increased the amount of time that waste or feces is transiting in the bowel). More time in transit means more time for toxin build-up, shifting gut in pH toward more an acidic environment, leading to overgrowth of non-commensal species (termed "bad bacteria"), as discussed in detail in Chapter 9.

LINK TO LIVER AND DETOXIFICATION

Significant magnesium deficiency can reduce vitamin D homeostasis. What is interesting is that clinical findings show that vitamin D is in a state of deficiency in states of high stress and inflammation. As discussed in Chapters 5 and 6, vitamin D is an important co-factor in the liver's detoxification pathways (Phase 1). The findings indicate that one of the mechanisms by which magnesium helps mitigate stress and inflammation is aiding in the absorption and activation of vitamin D, as substantiated in nutritional research.

Additionally, magnesium is an importation cofactor of glutathione synthase regulation – this is an important enzyme in the creation of glutathione, the body's most abundant innate antioxidant. Similar to vitamin D, glutathione plays a significant role in the liver's detoxification pathways.

Increasing magnesium levels can help aid in balancing pH, maintaining blood sugar levels, and supporting natural elimination of toxins through the liver, and serves as an important factor in bringing the body to balance in unlocking the skinny gene.

MINERALS WORK BEST TOGETHER

Nutritional research shows that magnesium and boron work synergistically. According to Gaby, low intake of magnesium contributed to low levels of boron, and Pizzorno's research has found that boron is essential for the absorption of magnesium. So, taking these minerals in combination will increase their capabilities to support a healthy metabolism. The greatest impact of their synergy is that they both are essential in the metabolism of vitamin D.

BORON AIDS DETOXIFICATION NEEDED FOR WEIGHT MANAGEMENT

Pizzorno's research, *Nothing Boring About Boron*, has shown that it also is a key nutrient in supporting detoxification pathways – it raises antioxidant levels to include glutathione. Additionally, boron has been shown by Pizzorno to decrease inflammatory markers and protect against pesticide-induced oxidative stress and heavy metal toxicity. These toxins all have the ability to create alterations, deletions, or substitutions via SNPs that can unravel our skinny gene design. As discussed in detail in previous chapters, supporting the body's ability to eliminate toxins is essential, for when it is dysregulated it can lead to weight gain. Boron appears to be an essential component to this metabolic regulation that is in reduced supply in our diet and therefore should be a priority in our skinny gene design.

DESIGN CHECK: HOW GOOD ARE YOU AT NOURISHING WITH KEY 'SKINNY GENE' MINERALS?

If your body starts giving you any of the following signs, then it is likely signaling that it is time to boost the cell environment and add in information and guidance via missing vital metabolism minerals.

Signs of a need for magnesium include:

- Eye twitches

- Brittle, soft, dry, weak, or thin nails; nails that split easily

- Itching, red patches, swelling, and blistering on your hands and feet

- Allergy-associated stuffy nose, runny nose, sneezing, and post-nasal drip

- Cluster or tension headaches

- Fatigue

- Low back pain

- Muscle cramps

- Cramps or uncomfortable sensation in the legs, especially while sleeping

- Painful periods or menstrual cramps

- Pain, redness, and swelling in various parts of the body, particularly the hands and feet

- Kidney stones

- Constipation and/ or diarrhea

- Memory impairment

- Irritability, especially around the menstrual cycle

- Sleep disturbances and/ or insomnia

- Weight gain

- Irregular heartbeat

- Poor carb metabolism

- Insulin resistance

- Cravings for chocolate

Signs of a need for boron include:

- Degenerative cartilage

- Brittle bones

- Memory loss

- Muscle pain

- Carpal tunnel syndrome

- Receding gums

- Sex hormonal deficiency/ low libido

- Vitamin D deficiency

How would the body respond if Liz added in the minerals that her body was telling her to boost?

Liz reached in the fridge to grab a fresh red apple, she set it on the butcher block to start slicing. As she positioned the knife in her hand, she noticed the strong white on her fingernails. She picked up her hand for a closer look and realized her nails looked manicured, no chips or cracks. Liz could not remember the last time her nails were not chipped and flaking. She was always struggling with strong nails and it seemed without trying they were back to normal. What changed?

Over the last month, Liz had started focusing on ensuring that she was getting key catalyst minerals to boost her metabolism. It was like icing on the cake for Liz, not only was she seeing the scale go down and her jeans were fitting looser, but now she finally had strong nails she had convinced herself that she was never destined to have. She didn't realize that the symptoms of weight gain and her weak nails were all correlated. Liz thought about it … could it be that all these years, her weak fingernails were just her body telling her that she was missing the minerals necessary to also keep her weight in balance?

The answer to Liz's question is yes. The body is always communicating what it needs and how you can better design your diet to "inform your form," but we just sometimes miss the body's clues, so when we pay attention and nourish with what our body needs, it gets back into balance.

YOUR SKINNY GENES DESIGN-KIT

Like Liz, are you missing optimal doses of magnesium and boron to spark the burn for your metabolism? Try adding in two superfoods packed with both boron and magnesium and give yourself seven days ... your body will let you know if this is a missing component that you need to add to your own skinny gene design.

Spirulina challenge

Spirulina is a blue-green alga and is believed to be one of the oldest life forms on Earth. Shauna Burton, in her article "How many bowls of anything does it take to get the nutrition of Spirulina?" gives the following summary on the benefits of this superfood:

"Spirulina gains its nutrients from the sun. The life force of Spirulina is chlorophyll. Chlorophyll absorbs the sun's nutrients, which powers the plant's life. When humans ingest chlorophyll, it enriches and oxygenates the blood. It is often called "green blood," because this form of chlorophyll is almost identical to human blood, except that its center is magnesium instead of iron. When chlorophyll is consumed, our bodies can replace the magnesium with iron, thus strengthening our own blood supply. Chlorophyll has also been shown to heighten the efficiency of all the organs and tissues of the body. It also contains DNA and RNA, both of which increase cell turnover and slow aging. It contains essential fatty acids (EFAs) to help improve cell wall structure and enable better cell-to-cell communication, which equals better body communication."

Most importantly, spirulina contains boron and magnesium. So for seven days, take ten grams of spirulina a day and observe any changes in your body's feedback related to low magnesium and/ or boron.

Exercise: An Apple a Day

Apples are a commonly overlooked superfood that has been shown to protect the body from nuclear fallout according to research captured in Ji's article "The Amazing Healing Properties of Apples." This research has found that apples have the capacity to remove "carcinogenic radioisotopes" accumulated in the body from nuclear fallout (i.e. near Chernobyl and Fukushima), as well as prevent the radioisotopes from absorbing in the gastrointestinal tract. Apples also are a source of boron and magnesium.

For the next seven days, your challenge will be to eat two apples a day, preferably throughout the day. This can be in a variety of forms: raw apple (whole, medium-sized), apple sliced and sauteed, apple puree (equivalent to one whole, medium-sized apple).

You can combine these two exercises or do them sequentially. The key is observing how you feel in terms of hunger/ cravings, energy, sleep, mood, stress. If you notice an improvement, explore with other foods rich in magnesium and boron or work with your nutritionist to determine optimal supplementation, so these minerals become a daily part of your nourishing design to unlock your skinny genes.

8

IGNITING THE FIRE

Isabella woke up again around 2 a.m. with raging heartburn. Even while in a sleepy state, as if on autopilot, she reached for her bottle of antiacids in her nightstand top drawer. The shake of the bottle had become a familiar sound and she could hardly remember sleeping through the night without it. In fact, she could recall needing tums since she was a child. She had joked with her husband that she should take out stock in the antiacid company. If anyone mentioned to her that they had never experienced heartburn in their life, she thought they must be lying. Deep down, she wondered if she would ever have a week when she wouldn't be reaching for them regularly. She had read online about side effects from long-term antiacid use, but she didn't know of any other solution that would prevent her from lying in bed in pain, with the raging fire below her throat. She couldn't help but wonder if there was something more going on ... was this somehow linked to her struggles with losing weight?

The stomach is the gatekeeper between what you consume and the highly precise workings of the intestinal tract. Therefore, it sets the stage for what you will take in (assimilation) and what you will remove (elimination). It initiates several feedback loops with other organ systems (i.e., the gastrointestinal tract and liver) to prime them for their job. Therefore, it is a key player in bringing balance to systems downstream, and as you will see in Chapter 9, a critical driver of unlocking our skinny genes.

Hydrochloric acid (HCL) plays a critical role in vitamin and mineral absorption and also helps reduce the toxin load by protecting the body from infectious agents such as parasites, bacteria, and other pathogens.

As we have previously discussed, the body uses acid and bases to bring balance across the system. This is no different in the stomach. As you may be aware, the most acidic environment in the body is the stomach. HCL has a pH of 1.5 to 3.5. If this is acidic enough to break down protein in meat, then one might wonder how it is that it doesn't destroy our stomach lining tissue. The answer is via a base, specifically sodium bicarbonate. Therefore, when we eat sodium bicarbonate is released first to protect the stomach lining and it is this base that triggers the release of stomach acid. In The pH Miracle, Young explains that "for each molecule of sodium bicarbonate made, a molecule of hydrochloric acid (HCL) is made." Therefore, the level of hydrochloric acid in the stomach is driven by the level of sodium bicarbonate released in response to food. Additionally, the body's level of sodium bicarbonate is dependent upon the body's stored alkaline reserves. Young reveals in the pH Miracle that the stomach pulls from the blood alkaline reserves to create sodium bicarbonate. This implies that the acid level in the stomach is dependent upon the alkaline stores in the body.

THE NATURAL STATE OF THE BODY (ACID-ALKALINE BALANCE)

The intestinal tract is lined with a single cell layer barrier defense that is primarily alkaline. Recall from Chapter 4, approximately 90 percent of the carbon dioxide that is produced from the cells in respiratory metabolism is used by the body to reabsorb alkaline minerals and make sodium bicarbonate for buffering gastrointestinal and metabolic acids. Also, as I explained in Chapter 5, the extracellular (interstitial) fluid is tightly regulated by buffers and maintained in a more alkaline state. As addressed above, the stomach acid is also driven by the body's alkaline level of the body. This shows that the body functions naturally in a greater state of alkalinity.

IF THERE IS NO FLAME, THE FIRE WILL NOT BURN

Now consider this: if the stomach is primarily an alkalinity organ and it pulls from the blood alkaline reserves to create sodium bicarbonate, then it is one of the first places we will receive feedback when there is an imbalance in acid/ base regulation in the body. Maintaining alkalinity of the blood, tissues, and organs are required to reduce dietary, metabolic, respiratory, and environmental acidic toxic waste build-up. In response, antioxidants play a key role in this regulation. Therefore, if your antioxidant status is low, then it means the system is too acidic. It is pulling from the reserves (across cells in the body) to bring balance.

When any of these systems are out of balance, there is a shift toward acidity. Increased acidity means less "alkaline juices" to draw from in the stomach. As we discussed above, less sodium bicarbonate equals less hydrochloric acid. As discussed in Liz Lipski's book, *Digestive Wellness*, some of the symptoms that can result from this are:

- Burping

- Fullness for an extended period of time after meals

- Bloating

- Poor appetite, especially in the morning

- Easily upset stomach

- Constipation

- Food Allergies

- Nausea after taking supplements

- Iron-deficiency anemia

Having sufficient stomach acid is essential to absorption of vitamins, minerals – especially trace minerals that function as critical catalysts in energy metabolism (as addressed in Chapter 7). Stomach acid levels determine the output of enzymes in the small intestine that drive the digestion of fatty acids, carbohydrates, and protein. This is because the body requires a strong acidic stimulus when food is exiting the stomach and moving into the small intestine, where the majority of food is digested. This surge in acid triggers a large output of digestive enzymes from the pancreas and bile from the gallbladder. As you will see in Chapter 9, bile acids, pancreatic enzymes, and hydrochloric acid are all required to maintain proper pH in the GI tract. If stomach acid is too low, then there is a decrease in enzymatic release and decreased ability for the small intestine to break down food molecules. This can show up as undigested food in the stool. When stomach acid is low, there is also a decreased release of lipase, an enzyme that aids in the digestion of fats and protection of the lining of the small intestine. Therefore, Low stomach acid can contribute to inflammation in the small intestine, and since it is the site of greatest nutrient absorption, it can impair the amount of nutrients getting into the body for utilization in metabolic processes. In *Digestive Wellness*, Lipski characterizes the symptoms that result from low stomach acid, leading to degraded functioning of the small intestine and inflammation, like the following:

- Abdominal cramps

- Indigestion 1-3 hours after eating

- Fatigue after eating

- Lower bowel gas

- Alternating constipation and/ or diarrhea

- Foul-smelling stool

- Dry, flaky skin and/ or dry, brittle hair

- Pain in left side under the rib cage or chronic stomach pain

- Acne

- Undigested food in stool

- Nausea

- Acid reflux/ heartburn

- Food allergies

LINK TO WEIGHT GAIN

Bile is essential to breaking down fat and making it easier to absorb and store for energy. When bile breaks down dietary fat, your body releases an enzyme that helps balance T3 and T4 hormones required to support healthy thyroid function – the driver of your basal metabolic rate (the number of calories required to keep your body functioning at rest)

Additionally, optimal levels of hydrochloric acid have been shown to be essential for the absorption of vitamin B3, niacin, in the intestinal tract. Vitamin B3 is an essential precursor to the glucose tolerance factor shown to regulate glucose metabolism. Therefore, low levels of stomach acid have a direct impact on glucose metabolism and can lead to insulin resistance and weight gain.

Low stomach acid creates a shift in the pH throughout the rest of the gastrointestinal tract. We know that an alteration in the pH will shift our body to higher energy harvesting microbiomes, leading to further impact on glucose metabolism, which is described in greater detail in Chapter 9.

Finally, the biggest correlation between low stomach acid and weight gain is through inflammation that arises as a result of altered pH in the gastrointestinal tract, stemming from low stomach acid. As mentioned above, when the pH isn't optimal, then enzymatic release gets impacted, leading to poor digestion of macromolecules in the small intestine, which creates irritation and inflammation in the villi that line the small intestine. The villi are comprised of fingerlike projections in the lining, which contains 6,000 to 25,000 villi per square inch of the intestine. That density is required to increase surface area for absorption of macro and micronutrients. When the large, undigested molecules become inflamed, the density of the villi decreases. Embedded with the villi in the mucosa of the intestine are lymphatic tissues designed as a first-line defense to protect from foreign invaders. When they are reduced due to inflammation, then the single-cell layer of intestinal tissue becomes susceptible to both pathogens and large molecules getting into the bloodstream. It is estimated that approximately 80 percent of our immune system is found along the intestine. When the intestine cannot effectively keep up the first line of defense, then the second line of defense is the liver. As you will learn in greater detail in Chapter 9, if the body is not able to absorb the required antioxidants and essential B vitamins to support the liver's added processing of antigens in response to pathogens and large, undigested food molecules, then it gets backed up in Phase I/ Phase II detoxification and the destruction of the "foreign" substance doesn't happen. This can overwhelm the immune system and create an allergic response, and in time create an overactive immune system, which can lead to systemic inflammation. Such a burden of toxins in the body is a direct cause of weight gain.

Your role is to become aware of what your body is telling you and if it is signaling low stomach acid, then it is time to replete your alkaline reserves through shifting the way you nourish toward your skinny genes.

DESIGN CHECK: HOW GOOD ARE YOU AT NOURISHING YOUR ACID-ALKALINE BALANCE?

If your body starts giving you any of the following signs, then it is likely signaling that your stomach acid is low and you may need to design an increase in your alkaline reserves.

- Burping

- Fullness for extended time after meals

- Bloating

- Heartburn

- Poor appetite, especially in the morning

- Stomach upsets easily

- History of constipation

- Known food allergies

- Nausea after taking supplements, especially in the a.m.

- Rosacea (red) on forehead, cheeks

- Weak, thin fingernails, easily chipped or bent

- Chronic allergies

Could it be so simple to resolve a lifelong struggle with heartburn?

Isabella never knew relief like finally solving her issue with heartburn. She realized that her years of taking antiacids had only fueled the problem, and it was such a pleasure for her to sleep through the night uninterrupted by shooting pains below her throat. She had always thought that she needed to decrease her stomach acid, but when she worked on increasing it naturally with apple cider vinegar at night and an AM cleanse, suddenly her heartburn subsided. Isabella was now reporting restful sleep, increased hunger in the morning, and regular bowel movements at least one time per day, and ease when buttoning her jeans. She finally could stop waking up her husband with a shaking antiacid bottle in the night and was proud that she never followed the notion to invest in antiacids, since she was seeing the impact from natural balancing remedies.

Contrary to popular belief, low stomach acid is the greater cause for commonly reported stomach issues. It is the stomach acid that creates a pressure vacuum to seal the esophageal sphincter. When stomach acid is low and the esophageal sphincter is not tightly sealed, symptoms of acid reflux can arise. What Isabella was experiencing was feedback from her body that her stomach acid was too low and this was likely contributing to downstream imbalances in her gut that were blocking her from reaching her weight goals.

YOUR SKINNY GENES DESIGN-KIT

Measuring Your Flame

The following two exercises are designed to help you explore your fire/ flame level needed to unlock nutrient absorption and facilitate toxin removal. In this exercise, you will utilize pH test strips to monitor your urine and saliva three times a day, to get a sense of your rhythms and patterns and to baseline your flame. Your test strips will indicate your acidity or alkalinity readings. After you have established your patterns over the course of a couple of days, you will layer in two practices to help further restore balance to your peak alkaline and acid periods, to enable the body to return to optimal function.

Exercise: Tracking your pH

For this test, you will need to purchase a urine/ saliva test strip kit, which can be found online, at drug stores, or provided by your nutritionist. To establish a true baseline of your body's level, plan to test over a seven- or ten-day period.

1) Test your urine pH. When you test your urine daily, you get to see the acid/ alkaline levels of your body.

When to test:

Take your urine at the following two points in the day:

1. First urination upon rising (after fasting). Since 2 a.m. is peak acidity (waste removal), this is going to show the status of your acid levels.

2. 2 p.m. Since 2 p.m. is peak alkalinity (alkaline reserves), this is going to show the status of your alkaline levels.

How to test:

1. Place strip in urine mid-stream for 2 seconds and shake off any excess liquid. Wait 15 seconds and match the color test pad to the chart to determine the rating.

2. Make note of any highly acidic meals and/ or highly alkaline meals and their impact on your daily pH.

Optimal range: The pH of the urine should run between 5.5 and 8.5 but ideally is around 7.0 - 7.2.

- 7.0 or higher: Indicates that your liver has adequate alkaline reserves.

- 6.5 - 6.75: You have alkaline reserves, but you need to replenish them.

- 5.75 - 6.25: Depletion of alkaline minerals.

- 5.5 or below: Indicates a very low (or no) electrolyte reserve. Your digestion and liver are likely affected by this.

2) Testing your saliva: When you test your saliva daily, you get to see corroborating evidence of the acid/ alkaline levels of your body.

When to test: You should take your saliva test first thing in the morning and also periodically throughout the day, testing about 2 hours after eating.

How to test:

1. Fill your mouth with saliva and swallow. Fill your mouth with saliva again and collect onto a spoon. Immerse colored test pad in fluid for 2 seconds. Wait 15 seconds and match to the pH test chart.

2. Make note of any highly acidic meals and/ or highly alkaline meals and impact on daily pH.

Optimal range: The pH of the saliva should run between 5.5 - 7.5 but ideally is around 6.7 - 7.0.

• 6.75 - 7.0: Optimal

• 6.25 - 6.75: Mineral and electrolyte depletion are moderate.

• 5.75 - 6.25: Depletion of alkaline minerals.

• 5.75 or below: Indicates that you may have a serious mineral and electrolyte depletion. You may get sore easily when exercising excessively. You need to restore your alkalinity reserves through nutritional rebuilding.

Exercise: Supporting the Energy Balance

Starting on day three of your pH test, start including the two practices below. Continue these practices throughout the remaining five days and make note of any impact on your pH test results

Since, the alkaline reserves are built from 3 a.m. to 2 p.m. (peak at 2). The acid levels rise throughout the night, primarily in the hours from 10 p.m. to 2 a.m. That matches with our body's period of cellular rejuvenation, repair, and "housekeeping." We can use easy nourishing tools to help support the body's balance through these steps outlined below.

Morning Cleanse (Alkaline Support)

- How to take: Take a.m. cleanse drink upon rising, but ideally between 6 and 7 a.m.

- Recipe: 1/3 cup distilled hot water, 1 tablespoon fresh lemon juice, and 1/4 teaspoon non-irradiated cream of tartar.

Nightly Apple Cider Vinegar (ACV) (Acid Support)

- How to take: Take diluted raw apple cider vinegar drink 5-10 minutes before your last meal, but ideally between 6 and 7 p.m.

- Recipe: 1 tablespoon raw apple cider vinegar in 3 tablespoons distilled water.

Remedy: Immediate Heartburn Relief

Any time that you have a heartburn flare, you can use the following recipe to balance the sodium bicarbonate levels and increase stomach acid to increase the pressure vacuum to seal the esophageal sphincter and decrease the "burn". The more frequently you rely on this remedy to calm your heartburn flare-ups, the more likely it will be that you won't need it any longer. When you use this in combination with paying attention to your body's feedback and acid/ alkaline balance and start working with the body to bring balance, you will find that heartburn will be an issue of the past, and it will serve you in impacting all the downstream systems associated with unlocking your skinny genes.

- How to take: Combine ingredients and take drink upon the first signs of heartburn.

- Recipe: 1/2 teaspoon non-irradiated cream of tartar combined with 1/2 teaspoon baking soda (aluminum-free) in 1/2 cup distilled water.

Based on this exploration, how balanced is your body's pH? Imagine how you can use this tool of self-evaluation to check in periodically and see how your levels have changed. You can try exploring when you are going through periods of high stress.

The key is observing how you feel in terms of hunger/ cravings, energy, sleep, mood, stress. If you notice an improvement, explore with new foods (see guidance in Chapter 10) to see the impact on your body's pH. This will allow you to design your favorite daily tools of acid-alkaline balance to unlock your skinny genes.

FANNING THE FLAME

Sarah came into my office complaining of unexplained weight gain. Her doctor had recently tested her for hypothyroidism but her results were negative. Sarah's was left still questioning her recent weight gain. Over the last nine months, she had slowly put on sixteen pounds. She had always been into weight training and had the goal of being at her physical peak by the time she hit forty (which was one year away). Despite all her efforts in the gym, starting cryotherapy, and IV infusions, she couldn't seem to get the scale to slide back down. When she considered her diet, she primarily ate a Mediterranean diet, 90 percent was homemade from fresh and organic or farm-to-table sources. She had a history of intermittent fasting that had always helped keep her lean and skinny, but she noticed that as she started putting on weight that she found herself eating more in the evenings, to the point that she felt like she couldn't sleep without eating. Specifically, she felt an insatiable craving for a little bit of ice cream before bed. Being a business owner, she had invested heavily in her professional wardrobe to always look "the part" and now for the first time in her life, she had rolls around her middle and her expensive clothes didn't fit well. She was now feeling even more insecure and this daily reminder of her weight was starting to affect her social and emotional health. She felt out of control over her body and so far from achieving her physical goals by forty.

What had suddenly changed in that nine months for Sarah, that had her body moving in a different direction from her intention, despite all her efforts to bring it back on track? As we can see in Sarah's case, there is something influencing her genetic expression and moving it toward expressing weight gain. If you recall from Chapter 3, a shift in our phenotype (physical appearance) is a result of alterations of our genes from new information. When this moves away from optimal expression (skinny genes), then it is usually a result of increased toxin load and too little nutrients. Nutrition is not just about what we consume, but what we are able to absorb (take in) in order to get to our cells to direct them (provide information) for assimilation. In our body's design, everything that comes in from the outside environment gets processed through two main organ systems. It is not surprising that these same systems play a significant role in determining what nutrients get absorbed (taken in) and in preventing roadblocks (toxins) from entering the body. These two systems in the body are the gastrointestinal tract (gut via the microbiome) and the liver.

ARE THE MAJORITY OF YOUR "SKINNY GENES" FOUND IN YOUR GUT?

The microbiome is comprised of an estimated one hundred *trillion* diverse microorganisms, weighing around two pounds, roughly equal to the weight of your brain. Compare this to the nineteen to twenty *thousand* human protein-encoding genes. It is clear that the most abundant system in the body, in terms of number of cells, actually isn't even human – it's bacterial and it plays a critical role in digestion. The profound number shows the importance that the body places on proper digestion and absorption.

Dr. Marco Ruggiero's twenty-plus years of research into the microbiome revealed that the microbiome is in essence an "operating system" for the body. He describes the following analogy: if your body is like a computer with memory, a hard drive, and processors, it will not function properly until you have installed an operating system (microbiome). He goes on to say that once the operating system is completely installed and fully intact, then the computer (body) can perform as it is intended. Similarly, it is only when the microbiome is completely installed over a period of time that the human body is able to reach its maximum human potential: integrated in our molecular signaling, metabolism, gene expression, behavior, etc. When the computer (body) starts shifting away from maximum operation (resulting in symptoms or dis-ease), just as an older computer gets inundated with internal and external inserts into the code that delay speed and performance, then the system needs a "reboot/reimage." The way to give the body a reboot is to restore a functioning microbiome.

As mentioned above, your body is made up of ten times as many microbes as human cells. Compare 20-23,000 human genes to 2-20 million microbial genes (e.g., bacteria, fungi, protozoa and viruses). This means 99 percent of the DNA operating in your body is in microbe cells not human cells. Over 10,000 microbial species have been shown to occupy various parts of the human body, with 7,000 different strains just our gastrointestinal tract alone.

We have co-evolved for millennia with the microbiome and the evolving research over the last five to ten years has demonstrated that an optimal microbiome ecology supports us in the following ways: It allows us to adapt to our environment faster than our own genes will allow us to adapt (responsive in three to seven days). It is involved in the maturation of the immune and protective functions (protection against pathogens and decreasing toxic load). It aids digestion. It supports metabolic functions – potentially greater metabolic capability, likely a role in directing metabolic activities. It facilitates elimination and detoxification. It has been shown to mediate disease pathophysiology (development and progression). It plays a significant role in the absorption and bioavailability of consumed nutrients and their metabolites. It is a major player in the gut-brain axis, with a significant role in production of neurotransmitters.

Specifically related to the metabolism, recent studies in the Frontiers of Microbiology (a peer-reviewed journal leading the research across the field of microbiology) have explored the relationship between the microbiome and energy harvesting, showing that our specific composition of microbiota works synergistically with the environment (to include pH) created along our intestinal tract lining, to drive harnessing energy from food. Specifically, researchers found that the genes with the microbiotic composition (various strains) encode enzymes that are essential for carbohydrate breakdown, nutrient acquisition, mineral release, energy harvesting, and guiding metabolic pathways.

In addition to aiding with the absorption of calories and micronutrients from food, the microbiome also produces important metabolic products, including short-chain fatty acids (SCFA), vitamin K, vitamin B12, biotin, folate, and thiamine (vitamin B1). The SCFA and enzymes improve the integrity of the intestinal lining, increasing its ability to remain selective in what passes through, as a strong barrier to prevent harmful bacteria, fungus, parasites, or undigested food molecules from passing through. One of the SCFA, called propionate, has been found to stimulate secretion of anorectic gut hormones, which are shown to suppress appetite. Through these actions, the microbiome has a significant role in reducing the toxin load from our external environment, thereby controlling inflammation in the body. The ability to support both nutrient assimilation and decrease toxic burden demonstrates how the microbiome is a power player in epigenetic control and the expression of our genes towards skinny genes.

It has been demonstrated through clinical observation that without a healthy microbiome, even if you eat what is considered a healthy diet full of vegetables and balanced protein and fats, this diet can become toxic through the creation of endotoxins (e.g., lipopolysaccharides [LPS]) that prevent absorption of the nutrients from the healthy foods you eat. In time, this can create a vicious cycle, where the toxins accumulate in the body, and in order for the body to prevent them from creating greater harm, it will pull antioxidants, minerals, and vitamins from other parts of the body. As a result, the nutrient-toxin balance in other parts of the body becomes disrupted, and leads to an even greater total body imbalance of depleted nutrients and a higher toxin load. As you will see below, this sets the stage for weight gain.

LINK TO WEIGHT GAIN

In the case of weight gain, let's explore the two major members (referred to as phyla) of the microbiota that shape the body's operating system, described by Dr. Ruggiero above: Bacteroides versus Firmicutes. Bacteroides is a genus of gram-negative bacteria. According to research, Bacteroides have a high capacity for breaking down proteins and carbohydrates and are considered a commensal (health promotive) species when in abundance. Firmicutes is a genus of gram-positive bacteria that, when in abundance, have been shown to have a negative influence on glucose and fat metabolism and are considered a harmful species when in abundance.

In order to understand the importance of these two phyla in optimal health, we just need to look at the model of the microbiome in breastmilk – the epitome of life-sustaining nourishment. Infants who consume only breastmilk create a gastrointestinal tract made up of greater than 90 percent Bifidobacterium, which highlights its importance from a biological standpoint, especially in immunity, health, and growth promotion.

In a recent interview, microbiologist and microbiome genetic sequencing expert Kiran Krishnan discussed the impact on metabolism of shifting between Bacteroides to Firmicutes. He raised study findings that a 1-2 percent fluctuation in the proportion of Bacteroides to Firmicutes can completely shift your metabolic fate.

For example, a human study was conducted to characterize the gut microbiota of people from many diverse locations around the world. The study found that those that presented with a lean physical appearance overall had less insulin sensitivity, less weight gain and obesity, and were less likely to have metabolic syndrome biomarkers. This same population of lean individuals had a higher Bacteroides and lower Firmicutes composition and tended to have a higher capacity for fat burn and a lower capacity for fat storage, as well as subjective reports of higher energy. The reason for this has to do with energy harvesting of the dominant microbiome family (phyla). This study showed that if you take two people – one person with higher Firmicutes and the other person with higher Bacteroides – and give them the exact same diet, the person with higher Firmicutes will absorb as much as 15 percent more calories from that same diet than the person with higher Bacteroides. The studies showed that this increase in calorie absorption can cause a 150-250 additional increase in calories per day. Every 150 calories taken in via the microbiome translates to one pound in weight gain over the course of three to four weeks and twelve to fifteen pounds of weight gain per year.

A NEED FOR REST

The body works tirelessly, with the aid of our microbiota, to maintain homeostasis balance. Just as the cells need an optimal environment to thrive, so does the microbiome. The driving factor for the types of bacteria (whether it is more inclined toward a population of Bacteroides or Firmicutes) is the pH in the gastrointestinal tract.

What we eat alters the types of bacteria that reside in our gut, which in turn changes the types of waste that are produced.

As we bring food into our body, the microbes in the microbiome get to work in concert with the liver's release of bile acids (see liver section below) and the stomach's release of hydrochloric acid and sodium bicarbonate (see Chapter 8). This activity in the gut (across the small and large intestine) is the process of digestion, where the food is broken down into component matter.

The issue is that we don't often give our gut the ability to rest and restore between digestion sessions. As we mentioned in Chapter 5, if you don't allow the body to clean and restore a balanced state, then it will tend toward an acid state. The result of this continuous breakdown process leads to bacteria dying off and byproducts (endotoxins, such as lipopolysaccharide [LPS]) accumulating in the gut and triggering inflammation. LPS is known to break the bond between the single layer of cells that make up the intestinal tract, providing a barrier to what gets into the bloodstream. LPS is actually used by researchers to induce inflammation in their subjects for testing. This accounts for the gas and bloating you can sometimes feel after a meal. If there is time for the system to regulate between meals, then in the space between the meal there is a natural "rinse cycle" that flushes out any of the buildups in the digestion process, to move it toward elimination. This is why it is critical to be able to rest and digest after a meal.

The importance in the body for periods of rest is exemplified in the connection between the parasympathetic system and the digestive tract through the vagus nerve, a nerve runs from the digestive tract to the brain and links to most major organs where it influences parasympathetic control. It is in this parasympathetic state where digestion is optimal. The vagus nerve, "reads" or is directed by the microbiome and the level of toxicity in the gut. Dr. Navaz Habib, author of the book *Activate Your Vagus Nerve*, highlights the correlation between LPS and epigenetic regulation via the microbiome in his report that the presence of LPS influences the gene that encodes nicotinic acetylcholine receptors – the critical receptor that balances the parasympathetic (rest and digest) versus sympathetic (fight or flight) response in the gastrointestinal tract.

Being able to allow the microbiome time to process

and break down food and restore the acid-alkaline balance in the gastrointestinal tract is one of the healthiest things you can do for your microbiome. It actually increases the diversity of the microbiota. In fact, in this state, studies have found increased gastrointestinal motility (i.e., improved movement of food from the mouth through the digestive tract until eliminated), increased nutrient absorption, and increased elimination and detoxification (waste removal). This is the state that will help you fully turn on your skinny genes.

GIVING THE GUT A LITTLE LOVE TO FURTHER HELP GET THINGS MOVING

A randomized control study showed that optimal ecology in microbiota of lactobacillus strains increases oxytocin production, which has a significant role in energy metabolism. Many are aware of oxytocin's role in bonding, connection, and calming. However, oxytocin has also been shown in animal and human studies to regulate gastrointestinal motility, inflammation, and permeability, as well as mucosal/ intestinal epithelium healing and repair. Research also shows oxytocin lowers blood levels of the stress hormone corticosterone and inhibits basal and stress-induced activity of the hypothalamus-pituitary-adrenal (HPA) axis. The HPA axis is linked to leptin (hormone that signals a feeling of being satisfied by food, full). Both elevated cortisol and leptin are associated with cravings, emotional eating, and weight gain. The hormone leptin helps to regulate appetite, making it easier to eat in moderation. However, when the body is in an inflamed state, it cannot hear the signal from leptin (reduced receptors to receive signal). High-fasting leptin levels are correlated with increased BMI and body fat percentage. Therefore, nourishing your microbiome toward higher lactobacillus strains will help it give love back to you through the release of oxytocin that further supports increased gastrointestinal motility, increased nutrient absorption, and increased elimination and detoxification (waste removal).

"75 percent of foods in the modern western diet are of little or no benefit to the microbiota."

— JAN CHOZEN BAYS, MD, *MINDFUL EATING*

The way to love your microbiome is to nourish it. Here are three easy steps to nourish your microbiome:

1. Remove processed grains, refined sugars, and processed simple carbohydrates. These are all highly acidic, with limited probiotic or prebiotic value to support the microbiome, and lead to increasing the

toxins (burden) that they must remove to sustain balance in the gastrointestinal tract.

2. Add in probiotic foods. Probiotic foods already contain an adequate dose of live microbes when ingested and therefore add beneficial bacteria to the gut. Examples of probiotic foods include kefir, cultured goats' milk, fermented vegetables, raw vinegars, kimchee, tempeh, sauerkraut, kombucha, cultured pickles, miso, natto, and olives.

3. Most importantly, add in prebiotic foods. Prebiotic foods are not broken down in the small stomach – they are fermented in the small intestine but not digested, so they can fully be utilized to feed the microbiome in the large intestine. Nourishing the microbiome in the large intestine with prebiotics helps them create and release organic (naturally produced) probiotics to balance the pH and flora composition throughout the rest of the GI. From an evolutionary and epigenetic perspective, these foods used to be a regular part of our diet but have decreased over the last several decades. Doctors such as Israel Dealba, a specialist in internal medicine and clinical professor, have reported findings that our ancestors used to consume about 135 grams of prebiotic fiber per day. Today, many Americans consume 1-4 grams per day. This is at best three percent of what our ancestors used to consume. Consider this when you think about what has

happened with our culture's metabolism over the last several decades.

Examples of prebiotic foods include raw leeks, raw/cooked onions, raw garlic, fennel, ginger, flaxseeds, seaweed, chicory root, jicama, Jerusalem artichokes, dandelion roots/ greens, asparagus, whole grains, oats, beans, bananas, radishes, tomatoes, carrots, burdock root, and psyllium husks. The goal is to consume ten to thirty grams per day, which has been shown to increase satiety and decrease appetite, support weight loss, and decrease gastrointestinal inflammation.

SKINNY GENES AND THE LIVER

Like the microbiome, the liver plays a significant role in supporting nutrient assimilation throughout the body and in eliminating toxins from entering the body. The liver and the microbiome work synergistically to balance the nutrient-toxin equation and maintain weight. This is outlined below.

There are two phases of natural detoxification (Phase I/ Phase II) in the liver for eliminating harmful toxins. These phases are dependent on the microbiome in the following three ways:

1. The liver is dependent upon the absorption and/ or creation of critical nutrients in the GI, received via the

hepatic portal vein, for direct uptake by the liver to facilitate these two phases.

2. The liver is also dependent upon the microbiome for keeping the integrity of the intestinal lining, to decrease toxins and pathogens from entering circulation, which can place a significant burden on the liver and lead to a backup in these phases.

3. The liver relies on the microbiome to balance energy harvesting to support balanced nutrient uptake by the cells, optimize insulin sensitivity, and decrease delivery of free fatty acids to the liver.

WHEN THINGS ARE GOING SMOOTHLY

Phase I is a pathway by which the body converts a toxin into a less harmful molecule. In order to achieve this, the body utilizes enzymes referred to as the cytochrome P450 enzyme group. The cytochrome P450 enzymes conduct various forms of chemical reactions (i.e., oxidation, reduction, and hydrolysis) to neutralize the toxins. Due to environmental influences (epigenetic), there are a number of SNPs (alterations) that can lead to back-ups in Phase I detoxification. What is important to understand is that these chemical reactions produce free radicals, which are unstable and highly reactive atoms. The free radicals can be harmful to cells and tissues when in excess, if not neutralized by antioxidants. If there are not enough nutrients or antioxidants to transition to Phase II, then the liver will package these toxins for delivery to adipose (fat) tissue.

- In order for this phase to be optimal, the following nutrients need to be absorbed from the GI: Vitamins B2, B3, B6, and B12; folic acid, glutathione, branched chain amino acids (leucine, isoleucine and valine), flavonoids (group of plant metabolites – see more in Chapter 11), phospholipids, vitamins A, C, D, and E; selenium, copper, zinc, manganese, CoQ10, thiols, and bioflavonoids. These nutrients are only available for the liver if the diet is rich in foods that provide them and with support of a healthy microbiome they are effectively absorbed.

- Vitamin D plays a critical role in supporting Phase I detoxification. As mentioned in Chapter 6, activated vitamin D has been shown to be essential for the expression of several Phase I biotransformation genes, as well as expression of antioxidant genes.

Phase II is the conjugation or "connection" pathway through which the liver attaches another substance (i.e., cysteine, glycine, or a sulfur molecule) to the neutralized toxin from Phase I, to make it water-soluble so it can then be safely excreted from the body (via urine or feces).

- In order for this phase to be optimal, the liver requires glycine, taurine, glutamine, N-acetyl-cysteine, cysteine, and methionine. These nutrients are only available for the liver if the diet is rich in foods that provide them and with support of a healthy microbiome they are effectively absorbed.

- Vitamin D plays a critical role in supporting Phase II detoxification. As mentioned in Chapter 6, activated vitamin D has been shown to be essential for the expression of several Phase II biotransformation genes, as well as expression of antioxidant genes.

The liver also contributes to the balance of the microbiome environment. The liver signals the control of the gallbladder's release of bile acids. Bile acids, pancreatic enzymes, and HCL are all required to maintain proper pH in the GI tract. Recall from Chapter 8 that this balance is driven by proper stomach acid levels to trigger a large output of digestive enzymes from the pancreas and bile from the gallbladder. When these three components are at optimal levels, then an environment is created in the gastrointestinal tract that is optimal for the growth of Bacteroides species to flourish and to prevent the overgrowth of pathogenic gram-negative bacteria – an underlying root of gut-liver inflammation.

WHEN THINGS ARE OUT OF BALANCE

Lack of nutrients, such as B-vitamins, combined with chronic inflammation from the poor microbiome flora and high lipopolysaccharides (LPS) will lead to liver dysfunction, diminished hormone production and blocked detoxification pathways.

Now consider this: if you are not getting enough oxygen and water each day to boost the activity of the mitochondria in the cell, then you are shifting the body toward insulin resistance. Recall from Chapter 4, insulin is "an anabolic hormone that promotes glucose uptake in the liver, skeletal muscle, and adipose tissue," according to Jonathan Temple an insulin resistance researcher. As we discussed, insulin resistance is triggered by excessive circulating glucose and results in decreased levels of glucose uptake across the body. The body compensates by increasing circulating levels of insulin (increases the signal). The rise in insulin increases the liver's creation of glucose storage in the liver (glycogenesis) and lipogenesis (increased fat stores). When this occurs over time, it impairs the liver's ability to perform some of its primary functions. Therefore, you are creating a cycle of directing the body to store fat.

Once this cycle of insulin resistance is established in the liver and skeletal muscle it shifts the signaling of metabolic hormones in the body. A state of insulin resistance, will drive the pancreas to increase insulin production to try to support cell uptake and reduce circulation glucose in the bloodstream. Also in this state of insulin resistance, leptin (considered by some to be the fat controller or weight regulator hormone) is decreased and therefore the signal to the brain to stop eating and that the body has adequate energy is hindered. Therefore, the individual continues to consume calories that further disrupted metabolic pathways. If not remedied, this will push the body beyond its adaptive processes, which can lead to increased weight gain (especially around the midsection) and in time can lead to the development of metabolic syndrome.

The challenge comes in when you want to lose weight, but the body is in this adapted state of storing fat. If you start cutting calories and increasing workouts, but never address the microbiome or the liver imbalances, here is what happens. With more workouts, you are breathing more and starting to increase the oxygen levels for your mitochondria (ATP powerhouse). Additionally, you trigger the metabolic pathway in the muscles to start taking in more insulin, again. As you do this over a period of time, you get to the point where your body can start drawing from fat stores to sustain the increased energy demand and fuel the cells.

However, as your body starts triggering the release from fat cells, what is sitting inside those cells? The toxins that the liver couldn't process previously. You unwittingly create a flood of pro-oxidants, imposing a greater burden on the liver. Therefore, the liver demands more cofactors (minerals, B-vitamins, vitamin C, A, E, D; amino acids, etc.) to enable the detoxification process. The problem is your reduced calories and microbiome composition do not facilitate the higher demand. Therefore, if the liver can't safely remove the toxins, then they will go back into fat stores. What's more, the liver will send signals to other systems in the body to help balance the demand:

- Thyroid: Thyroid hormones control the rate at which calories are burned. The liver regulates the balance of thyroid hormones. If the liver is in an impaired state, it decreases activation of the hormone T4 to T3 (active form) and sends a signal to the thyroid to decrease the burn rate through lowering thyroid hormone production. The liver requires T3 to store glycogen and without it, blood sugar regulation becomes a challenge, and blood sugar regulation issues become apparent and can present as hypoglycemia. This can show up in the body as shakiness, dizziness, sweating, hunger, fast heartbeat, inability to concentrate, and/or confusion between meals.

- Hypothalamus: The hypothalamus controls the body's appetite and energy homeostasis. The hypothalamus measures the amount of energy stored in the body fat. In order to reduce the release of body fat into circulation (so as not to overwhelm the liver), the liver will signal to the hypothalamus to shift the metabolic response to exercise. This means that even though you may increase your amount of effort in exercising, your body will lower the effectiveness through your rate of energy output. So you may be working harder and increasing the intensity or frequency of your workouts but your body is reducing the resulting burn rate.

- Estrogen Dominance: Part of the liver's role is to regulate the balance of sex. It transforms or removes any excess of these hormones from the body through the Phase I/ Phase II detoxification pathway, which is then excreted through the bile. If the liver does not have enough glycogen stores, then it cannot produce glucuronic acid, which is used in Phase II detoxification of estrogen. If Phase I/ Phase II pathways are backed up and/ or bile is not flowing, then estrogen can build up in the tissues and block the release of the thyroid hormone, further contributing to high levels of estrogen in the body, termed estrogen dominance. Estrogen dominance, high levels of estrogen, is associated with around the belly, butt, thighs, and hips.

• Cortisol and Adrenal Fatigue: Part of the liver's role is to regulate the balance of cortisone and other adrenal hormones. If there is low-circulating T3, due to the liver-thyroid feedback loop mentioned above, and the body is in a hypothyroid state, then the body compensates by over-activation of the adrenal glands to overproduce cortisol. The issue is that excessive cortisol also inhibits the conversion of thyroid hormones T4 to T3. However, if the liver cannot balance cortisol levels, then cortisol can build up in the body and create a vicious cycle, further depleting the adrenal glands – which has been associated with increased cravings, emotional eating, insulin resistance, and weight gain.
• Added Insults: Insults such as toxic environmental exposures (chemicals, pesticides, etc.), artificial sweeteners, alcohol, non-steroidal anti-inflammatory drugs (NSAIDs), and medications will only continue to stress the liver's natural pathways by increasing the toxic load that must be processed and reducing the antioxidant status within the liver. If this is not balanced with all the required nutrients and amply antioxidants to keep these pathways open and properly elimi-

nating, then the liver will package the excess toxins away in adipose (fat) cells.

This is where you may feel like your body is working against you. However, all these mechanisms above are depicting adaptive pathways the liver has shifted to in order to preserve and protect your tissues from stored toxins. This shows the many ways your liver is working to support you based on the environment and resources (nutrients or lack thereof) that it has available to perform its basic functions. It is up to you to design your diet so that you ensure intake and absorption of all the required nutrients and amply antioxidants to keep these detoxification pathways and ensure the ready removal of stored body fat.

These alternative pathways in the liver due to high toxin load accounts for why people say, "In my twenties, it was so easy to lose weight," and "I am getting older, so it is harder to lose weight." It is more a matter of cumulative insults over time that have shifted the body into a state of adaptation that is hindering it from its normal – or what is deemed "youthful" – operating system. Once you restore the balance and nourish fully, then the ease of maintaining weight is restored. One of the foundations to getting to that maintained weight is nourishing for your microbiome (to keep things moving) and liver (support daily detoxification) daily.

ADDITIONAL EVIDENCE FOR AN EPIGENETIC NEED FOR REST

Researchers have found that in the case of insulin resistance leading to weight gain that peroxisome proliferator-activated receptors (PPAR) found in the liver were downregulated (decreased due to epigenetic influences). In comparison, the researchers found in lean individuals PPAR were upregulated (increased expression). The same study concluded that PPAR was in transcriptional control, meaning directly activating/repressing gene expression of genes regulating glucose transport and insulin sensitivity, lipid metabolism, oxidative stress, and inflammation. This study shows that within the liver you have a natural tool of epigenetics to switch your body design towards one of leaner composition. The natural question then would be how might you switch this on (keep upregulated) and keep it on? Multiple studies have shown that PPAR is most activated in periods of fasting. This is another example of how the body reveals its need for rest. Knowing that the microbiome and the liver work synergistically to support the nutrient-toxin burden on the body and both demonstrate a need for rest (fasting) this should be considered a critical component in designing your skinny gene diet.

DESIGN CHECK: HOW GOOD ARE YOU AT NOURISHING YOUR MICROBIOME-LIVER BALANCE?

One of the primary ways your body reveals the status of your microbiome-liver status is through your gastrointestinal motility (bowel movement frequency) and your bowel movements type (see more detail are provided in the Design exercises listed below). Although it may seem off-putting to some to pay attention to and observe their own bowel movements this is valuable feedback from the body provided daily and optimally a couple times a day to inform you of your status. As described above, understanding this status so you can better design your diet is an essential component of unlocking your skinny genes.

What if Sarah's sudden weight gain was all from a shift in her microbiome-liver balance?

In order to get to the bottom of her unexplained weight gain, Sarah had a functional microbiome analysis via whole genome sequencing to determine the current blueprint of her microbiome ecology and the presence of any pathogens that were causing dysregulation. What was found is that due to a shift in her stress level and decreased intermittent fasting (which her body had become accustomed to), she had inadvertently shifted the abundance of bacteria that keep pathogens such as Bilophila wadsworthia in check. Bilophila wadsworthia is pathogen that directly challenges the microbiome-liver pathway and is associated with significant changes in fat and glucose metabolism. Specifically, Bilophila wadsworthia upregulates (directs an increase) in the global synthesis of LPS and downregulates (decreases production) of SCFA required to calm mucosal inflammation and to keep the intestine impermeable, allowing increased permeability of bacterial endotoxin LPS, which burdens the liver. Bilophila wadsworthia has also been shown to modulate genes involved in bile acid homeostasis. Recall, bile acids play a supportive role in regulating metabolism through supporting balanced pH within the GI and triggering the release of pancreatic enzymes such as insulin. The Bilophila wadsworthia present in Sarah's GI lead to a change in her GI pH which also made her susceptible to yeast overgrowth. This explained her daily craving for ice cream. The combination of Sarah's increased sugar and dairy consumption, decreased fasting, presence of a pathogen, shift in GI pH, and yeast overgrowth set the stage for her unexplained weight gain. As a result of these findings, Sarah began a dedicated strategy to balance her microbiome to naturally eradicate Bilophila wadsworthia, calm down yeast overgrowth, and increase vitamin and mineral uptake to support the liver and improve bile production. Within three months, Sarah was able to shed fourteen of her sixteen-pound weight gain. Additionally, her doctor confirmed that her thyroid hormones had returned to optimal range. Her cravings seemed to disappear overnight. Most importantly for her, she was fitting easily back into her professional wardrobe and back to feeling full confidence in the office.

YOUR SKINNY GENES DESIGN-KIT

Like Sarah, knowing the status of the microbiome can save you weeks, months, even years of frustrating weight gain. Working with a nutritionist specializing in whole-genome microbiome functional analysis you can get a comprehensive, "high-resolution" picture of your microbiome ecology, and thereby you can understand its specific impact on your liver-thyroid-adrenal function, which challenges your metabolism. However, there are also simple exercises you can do at home to start shaping the picture of your status, that may be contributing to shifting you away from expressing your skinny genes, outlined below.

Exercise: Moving Right Along

Step 1: Conduct a GI transit test. This test is designed to get a real-time assessment of the status of your gut motility and vagus nerve activation (are you in parasympathetic "relax and digest" part of the day or are you predominantly in sympathetic "fight or flight").

Ideally, you will do this on day 1 and day 7 of this exercise. If you haven't completed your first transit test by day 7, wait to start the second test until it is complete, giving 1 day in between. When you have completed both tests, compare results.

• Put a spoonful (about 1 tablespoon) of white sesame seeds in a glass of filtered water and drink it down, without chewing the seeds. Our body cannot digest the covering of the seeds, so they will show up in the stool. Note, if you have a known sesame seed allergy, you can use activated charcoal – two capsules – alternatively. This will show up as black stool.

• Watch and track. What you are looking for is the time it takes to see the first seeds in our stool and then when we see the last seeds to come out in our stool.

Track the following:

1. Date/ time started transit test

2. Date/ time first observed white sesame seeds/ activated charcoal in stool

3. Date/ time last observed white sesame seeds/ activated charcoal in stool

Evaluating the transit. Optimal window for first observed is 12-20 hours. Seeing the first observed window at 16 hours is aligned with optimal digestion and motility function.

Step 2: Nourish your gut-liver connection. Using the list of prebiotic foods above, add in 10-30 grams, equivalent to 3/4-2 tablespoons of prebiotic foods per day, to your skinny genes design for 7 days. A study found that significant changes can be observed in the microbiome within 3-7 days. For more on how to use food to balance the microbiome for sustained change over time, see the 3:1 rule in Chapter 10.

Step 3: Rest. As we learned above, one of the biggest impacts on optimal flora and liver health is time to rest, digest, and eliminate. So during the next seven days, you will be building in progressive periods of rest to increase time between eating and extend the period of repair and elimination during night and early morning hours.

- Day 1 and 2: Set a time to stop eating in the evening, ideally after dinner, and calculate an eight-hour window. Commit to not eating during this period.

- Day 3 and 4: Calculate a nine-hour window from your last meal to your first meal the next day. Commit to not eating during this period.

- Day 5-7: Calculate a ten-hour window from your last meal to your first meal the next day. Commit to not eating during this period. Additionally, during that day, allow for 2 hours before and after any meal to eat (four-hour window). Incorporate "breath snacks" (see Chapter 4) and proper hydration throughout the day.

Step 4: Watch. The beauty of the body is that it is always giving you feedback on its current state. No more true is this than in the evaluation of our bowel movements. Again, this is feedback that most people overlook. For the next seven days, you are going to get comfortable and familiar with looking at what your bowel movements have to say about the status of your microbiome and gut-liver connection. Throughout the day, track the following information.

- Number of bowel movements during the day.

- For each bowel movement, what is the type (select from one to seven below, based on standard Bristol stool chart)?

1. Type 1: Separate hard lumps, like nuts; hard to pass

2. Type 2: Sausage shaped, but lumpy

3. Type 3: Like a sausage but with cracks on the surface

4. Type 4: Like a sausage or snake, smooth and soft

5. Type 5: Soft blobs with clear-cut edges

6. Type 6: Fluffy pieces with ragged edges, mushy stool

7. Type 7: Watery, no solid pieces, entirely liquid

Step 5: Reevaluate your transit time. See guidance in Step 1 above.

Your ability to eliminate and keep things moving through the bowels opens up the pathway for the liver to support toxin elimination. Monitoring your transit time and daily BMs can be a body feedback to inform you on how well your body is doing at assimilating nourishment and eliminating toxins through the microbiome-liver pathway. Understanding the status of this pathway and working to optimize it can be a significant influencer in keeping your skinny genes turned on.

As you can see, when it comes to the microbiome, it is all about the composition that has the greatest impact on both the nutrients you are able to absorb and your ability to eliminate waste. By shaping the composition of the microbiome, you change the quantity and quality of nutrients absorbed to be utilized in the body. This is a more powerful factor than changing the nutrition alone. When the microbiome is doing the heavy lifting in helping keep out toxins and keeping them moving through to elimination, then the liver can focus on safely eliminating the toxins in the body to include those packaged away in fat cells. Therefore, nourishing the microbiome with a few tablespoons a day and lots of rest can be an essential part of your skinny gene design.

FUELING THE FIRE

Fiona walked up to place her lunch order. She stared and stared at the menu, scanning over all the dishes. Which one would be the healthiest for her? Which one wouldn't upset her stomach later and leave her full of regret? What was the one that would make her work friends the least judgmental? She felt confused and overwhelmed by what should be such a simple choice. Looking up, she saw the cashier waiting patiently for her order. She looked over to her friend who had just ordered ahead of her, and said, "I'll have what she's having." Thinking at least this way, she knows she won't be judged, and she will just have to roll the dice and see how it treats her.… but she couldn't help thinking "There has to be a better way".

Food nourishes the spark of life within the body. If you think about it, the same elements that we need to thrive – the sun, the soil (minerals), water, and air – are what is required to make the food that we eat. As we have seen across the previous chapters, nutrients have a tremendous potential to guide the expression of our genes to the fullest potential.

> *"Food is a language that speaks to the genes."*
>
> — JEFFERY BLAND

If this is possible, then why are we currently seeing a period of great obesity in our history, when food is most abundant?

COPIOUS AND YET ... CONFUSED

At no point in our history have we had more access to food, and at the same time been so confused about what we need to eat. How connected are we really to what we eat? Many people these days see food as a "means to an end" and in time, this reinforces the belief that all calories are the same. Therefore, if you have a brownie (as long as you only have half) and that cup of coffee, you are doing your best to make sure you don't fall asleep in your afternoon meeting. Based on what you have learned thus far about navigating your body, how do you think your body feels about that choice? Or maybe if you are like Fiona, you didn't fully make the choice and instead let someone else choose for you. How much work, effort, and energy is it going to take for your body to extract the nutrients it needs to keep you up and awake, and all your metabolic pathways running optimally, when you eat that brownie and coffee? If your body was in control would it make the same choice?

What is the relationship you have with your body? Take a minute and think about this. If you nourish your body, based on what you truly believe about it ... what is the *real* relationship you have with your body?

When you consider our confusion with food, could it arise because we aren't aware of the messages we are sending to our body and how it is taking those messages and shaping itself? Could it be that our block to our skinny genes is a matter of miscommunication?

WHAT ARE YOU REALLY CRAVING?

Take for example a child. If you bring them to a farmer's market often throughout the year and encourage their exploration of various foods that they are drawn to, there is something amazing that happens. In the late fall, they might be drawn toward dark, earthy mushrooms that are immune boosters and naturally high in B-vitamins and vitamin D to help support the transition in the seasons. In the summer, they may be drawn mostly to the bright-colored fruits that are bursting with polyphenols, antioxidants and enzymes that support digestion and elimination. These examples highlight the raw design talent we have when we live more in our bodies and let it drive us toward foods.

As we get older, somewhere along the way it appears that this signaling gets muted and distorted. Cravings often are a misinterpretation of an alarm – a warning of what is not serving us and leading to symptoms or dis-ease. Dr. Liz Lipski, author of the book *Digestive Wellness*, points out that many foods (e.g., wheat and dairy) produce protein molecules that mimic our natural endorphins, which explains why we can temporarily feel better when we eat them and continue to crave more. However, if we are numb or disconnected from our body, we don't make the connection to the symptoms (outside the window of pleasure) when the mimicked endorphins wear off, which then drives the body to express greater symptoms of imbalance.

Now, you may ask yourself, "What if my body is driving me toward carbohydrates?" This isn't a matter of willpower or a flawed system. Take, for example, something that on a logical level you know doesn't serve your body (e.g., a Krispy Kreme donut), but that you crave. When that food is available, you eat it and have a hard time stopping at just one. Or sometimes, you find that you even go out of your way just to get it. You eat it and later feel horrible in both mind (guilt/shame) and body (bloated, indigestion, fatigue). This craving is simply pointing you toward an underlying imbalance, the same way that the child at the farmer's market was drawn to mushrooms to get more immune-supporting nutrients that he may be lacking in the winter months. However, your cravings for poor quality food are a way for the body to call your attention to the foods that hurt you the most, so you can be aware of the imbalance. If we don't connect with the experience to look more closely and use the feedback from the body, we will miss the opportunity and simply continue to repeat the pattern. Again, in the body's wisdom and its capacity to always working to bring balance, it will find an even "louder way" to draw your attention to what you may have missed.

Here is yet another way that food can be a powerful tool guiding you toward an expanded awareness of what you need. As we mentioned in Chapter 5, water can imprint frequencies. Sometimes these frequencies can get trapped in the body. As we start working to bring balance to the body, this can trigger a signal that we are ready to release this trapped emotion. If there is an experience of that emotion being linked with food, then as this starts coming up, we can experience a deep craving for a particular food.

For example, take the case of my client Chloe, who was trying to get back in shape. She finally was in a great place in her life. She had just been promoted at work. She had a great brand new apartment, and with her new income, had no trouble filling her new place in a way that felt perfect for her. Dedicating so many hours at work had resulted in a less-than-optimal lifestyle and she found herself about twenty pounds from her ideal weight. As she started trying to shift her lifestyle and started working out, suddenly she found a repetitive craving for Fritos corn chips. It seemed like she would do so well during the week, but by the weekend, she would find herself arm-deep at the bottom of a Fritos corn chips bag. She knew this wasn't serving her weight loss goals, but she couldn't seem to shake the craving. After working with Chloe on mindful eating practices, she was finally able to link the Fritos corn chip craving to an emotion. Her brother had passed away tragically in a car accident when he was a teenager. She had never fully grieved his loss and now that she was in a secure place in her life and her needs were met, old and trapped emotions were coming to light. The last memory she had with her brother was a weekend movie binge-fest, in which they ate endless amounts of Fritos corn chips. This food craving was signaling to her a deep-rooted emotion that was no longer serving her and was a roadblock to her weight loss. It was not the craving that was hindering Chloe's success, it was what was behind the craving that was the key to unlocking and turning back on her skinny genes.

This shows that cravings can also signal what we need on the emotional level as well. When we become aware of feeling into what we need and allowing the information to rise up to meet our mind, as opposed to our mind trying to force what it "thinks is best," then we make choices that are more fine-tuned to nourish on all levels.

MOVING INTO A WORLD OF DESIGN

When you think of a designer, what characteristics do you see? Are they rigid and restrictive? Or are they resourceful and inspired? The reason I like to think about building a diet to fit your life in terms of designing is that it is all about creativity, about being open, about what you are drawn to. This creative connection with food is innate ... but over time, like any tool, it can get dull.

If nourishment is a tool for honing your physical form through the upregulation/ downregulation of genes, then it is of the utmost importance that you are aware of what you consume, since it literally becomes you. As we have discussed throughout previous chapters, the way you nourish will either facilitate or block the pathways toward expressing your skinniest genes.

When it comes to food, it is not about quantity but about quality. This also is supported by what we see occurring in the body when we diet. The biggest issue with restricting calories without dedicated attention to the variety and quality of nutrients is that if you just reduce intake then you often end up consuming an incomplete profile of nutrients. As discussed previously, if you don't have abundant nutrients such as B vitamins or minerals as cofactors, then you are unable to break down the food into energy. At a certain threshold, you can dip below the energy levels needed to maintain basic body functions and to fuel the cell (see mitochondria and energy discussion in Chapter 4). This signals to the body that it is in a state of starvation. The thyroid is often most affected by the frequent dieting and it shifts its release of thyroid hormones, which signals to the liver to reduce thyroid hormone activation and increase fat storage. Additionally, the increase in fat storage signals via leptin secretion to the hypothalamus that the body is storing fat and shifts the set-point for satiety. Additionally, the hypothalamus will shift the body's basal metabolic rate, meaning that if you increase working out, the same level of input now will burn fewer calories to conserve energy. This cycle makes it harder to lose weight and creates the yo-yo dieting syndrome.

ENERGY BEYOND BUILDING BLOCKS

In order to understand the powerful information that food provides to the body, you must consider the body in terms of frequency. In 1992, Bruce Tainio reportedly built the first frequency monitor in partnership with Tainio Technology and Eastern State University in Cheney, Washington. Bruce found that a healthy body frequency is 62-72 Hertz and that the lower this frequency becomes, the greater the report of illness and dis-ease. This research was highlighted in Wilfred

Campbell's article "Vibrations: Vibrational Frequencies and Food," where additional correlations between low frequency and disease are presented. Campbell found that fresh foods and herbs had a frequency of 20-27 Hertz and even higher if grown organically and eaten freshly picked. However, processed and canned foods were found to have a frequency of 0 Hertz. As we discussed in Chapter 5, frequency drives pH in the body, which drives our biochemistry and epigenetic regulation of our genetic expression. This implies if we want the greatest outcome from the food that we eat to inform our optimal genetic expression, such as our skinny genes, then we need to ensure we are getting in more of our food that is in a form of the highest frequency and decrease the intake of processed, low-energy foods.

What if Fiona started seeing food as a form of self-care?

During a nutrition consultation, Fiona was venting about feeling like she either didn't know what to eat or she had no control over how much she was eating. She had just reviewed the findings of a recent pH testing assessment and functional lab results. They revealed that she had a high toxin load and was not taking in enough nutrients to balance the equation. When talking through changes to bring the equation back into balance, she revealed it made sense to her logically, but emotionally she felt frustrated, angry, and defeated.

In order to get at the root of her feelings, she was asked if she could picture her optimal body and describe what it feels like to be in that body. Fiona described that it felt light and happy. When asked how it felt like to be in her current body, she described it felt heavy, tired, and angry. She then explored the concept that both of these versions of herself, the ideal one and the current one, were just two different physical expressions of her. The difference was that one was lighter, higher energy, and filled with nutrients to keep her in balance. The other was heavier and burdened by toxins. She was then asked to think of a food that she felt she couldn't stop eating. For her, it was donuts. When she was asked to feel from her body's perspective which version of herself did that food fuel, she realized it clearly was the heavier, tired, angry version. Fiona said it was then that it clicked. She was choosing this form, this figure that she was expressing ... and the beauty was that this meant that she could also choose the other "lighter" form, too. What her figure expressed truly was up to her, she just had to get practice in feeling what it was like to nourish her "lighter" form ... and in no time, her body would reflect on the outside what she was building toward with each and every breath, drink, and bite.

So the question is, how to get quality, nutrient-dense food? A 2009 large study found that 80 percent of Americans are low in virtually every color category of fruits and vegetables, leading to nutrient deficiencies in plant-based vitamins, minerals, and phytonutrients. As you learned in the previous chapters, the foundation of balancing nutrient to toxin ratio is found is high alkalinity, minerals, vitamins, and phytonutrients. These are found in abundance in fresh, organic fruits and vegetables that our genes have evolved eating in order to thrive.

YOUR SKINNY GENES DESIGN-KIT

Sharpen your skinny gene design skills by committing to exploring the following ways to nourish your body with food. Dedicate 7-10 days utilizing at least one of these practices and observe your body's response.

Start with "Eating the Rainbow"

An easy way to do this is to let color be your guide and "eat the rainbow." Specifically, phytonutrients work in the human body to stimulate enzymes (for gene regulation) and are referred to as "nature's biological response modifiers." They help the body get rid of toxins, boost the immune system, improve cardiovascular health, promote healthy estrogen metabolism, and stimulate the death of cancer cells. Phytonutrients in food come in all different colors – green, yellow-orange, red, blue-purple, and white. The goal would be to layer into your design all the colors of the rainbow, through fruits and vegetables, throughout the day.

A Shot of Goodness

Drinking your vegetables can be a good way to get a head start in your daily nourishing design. As long as you are getting a good amount of fiber in your diet, then vegetable juices provide a high concentration of rapidly useable alkaline salts, vitamins, minerals, active enzymes, and chlorophyll. Especially in the first half of the day (before two p.m.), vegetable juices will help to increase your alkaline reserve throughout the day. The following recipe is adapted from Robert O. Young, author of the book *The pH Miracle*. Make a vegetable juice with cabbage, celery, carrot, beet and/or cucumber. Start small, with a fourth-cup and build up to one full eight-ounce cup, since vegetable juices can be a bowel cleanser (if so, likely needed). Dilute with distilled or reverse osmosis water, in a 1:10 dilution. You can add a fourth-cup of fresh-squeezed lemon juice to water before adding juice, to increase alkalinity and cleansing of vegetable juice.

Rule of 1:3

Since we all live in the real world, it may not be possible for you to fulfill your design goals each and every meal. However, it can take the body a few meals to recover from one poor quality meal. For example, increased sugar intake decreases available B-vitamins. If there are not B-vitamins available in circulation, then the body will pull from other areas of the body (i.e., the liver). If these B-vitamins are not repleted (replaced), then that is going to lead to an impairment in the function of that organ. In the case of the liver, it will decrease the liver's ability to detox. As we discussed in Chapter 9, any backup in detoxification will increase the likelihood of fat storage (if toxins can't be processed, they are stored away safely in adipose [fat] tissues). However, the body will always find its way back to balance if you ensure that you:

1. Replenish what was lost,

2. Nourish for the moment, and

3. Store for the future.

To do this, you need to follow the 1:3 rule. This rule was established by Dr. Oscar Coetzee, a nutritionist who has an established history of helping clients reverse metabolic disorders. For every *one* unhealthy meal (e.g., highly acidic, low-energy, highly processed/refined) you consume, you owe your body *three* optimal meals in a row. Build on your previous experiments to design your perfect plate.

Designing a diet to get the most out of the food you consume only works when you are nourishing on all levels of the body to ensure the balance of enzymes, cofactors, energy, etc. to run the metabolic pathways optimally. Following the steps outlined in the previous chapters will help pave the way. As you build in balanced nourishment, the blocks to weight loss start to resolve and the feedback from the body becomes clearer. It is from this place that your designer diet can open to abundance in trying new foods and relearning what makes your body thrive, hum, and feel alive. It is in this place that skinny genes are unlocked and stay turned on for the long term. Determining what this design is that creates the expression of your skinny genes is about your daily choices and exploration. Taking a lesson from Fiona's experience, the rule of designing is to make it fun, mix it up, and be inspired ... and always observe the body's feedback.

ON THE WAY TO SKINNY ME

When you get more connected to your body, you can move away from the templates and move toward personal knowledge, knowing all day, every day exactly how to nourish to express your skinny genes!

DESIGNING TAKES TIME AND INSPIRATION

Using the template provided in this book will solidify the relationship that you have with your body, by empowering you to give it the much-needed time and attention it deserves. When you take the time to observe and feel into the body, you will reclaim your expertise in knowing what it needs to rebalance … that is where your skinny genes are waiting.

LIKE RIDING A BIKE

If you are like most people, you likely have spent much of your life "eating from your head" or eating what someone else put on your plate (e.g., parents, partners, etc.) and taking on their pattern and beliefs about what was right to put in your body. Therefore, the concept of eating from the body has been lost. We are not taught, like the ancients, how to use nutrition to shape the best expression of genes. These ancient families were able to "hit the ground running." We are not there any longer. However, regardless of your age at the start of this journey toward getting re-acquainted with your body and shaping of your genes through nutrition, it is what you are made to do. It's literally and figuratively *in your genes*.

Give yourself time for exploration, allow it to feel awkward. Think of the first time you tried to ride a bike. You may have gotten frustrated and it may have felt all wrong, but in time you learned the art of balance. You learned to navigate the feedback from the bike and the road. In time, you were able to push harder and go faster. This is no different. Practice, time, and attention are all you need. Like riding a bike, once you learn it and connect with the feedback/responsiveness system, you can't forget.

NAVIGATING THE NEW DESIGN IN YOUR OLD ENVIRONMENT, SOLIDIFYING YOUR PERFECT FIT

Through this process of exploration and discovery, you will see that you no longer have to guess what you need and how to bring balance to the body. A balanced body doesn't hold weight (it is an energy stealer). Being able to be the builder of the tailor-made diet to optimize your skinny genes restores power in your choices for how to nourish in a way to serve your body. If you get stuck, then all you need to do is go back to the template and tap into the body to see where there is an imbalance. The more you do this, the less the pendulum will swing, until you can get to the point of knowing immediately what to do when an imbalance is signaled, to restore harmony. It is from this place of knowing that choices become abundant and skinny genes stay on!

THE ALTERNATIVE

You are the only one with the secret code to your skinny genes. Therefore, if you don't take the opportunity to unlock this code, and turn to everyone else to learn their code (what has worked for them), you could spend a lifetime spinning in circles on what and how to eat. This means years of unmet weight loss goals, increased frustration, worry or fear that something is wrong, greater health issues, gaining more weight back now or in the future, and feeling out of control of the body, with nowhere to turn. My guess is that if you get to this point … you may start realizing that the answers waiting within are worth the effort of finding.

So, why wait? Why put it off? You have nothing to lose but pounds.

What this process has set up for you is a daily template to get into the body (and out of the head) on a regular basis throughout the day. Through this process, you know what the body needs most to fuel metabolism, decrease stress, increase detoxification, and increase energy are the following: The most essential nutrient for the body is oxygen – deep, cell-supporting, ATP-boosting oxygen. Next, the body needs water to keep it thriving, flowing, and clean (mind and body). In order to prime for the burn the body needs ample supply of vitamin D. In order to unlock the vitamin D and spark the flame the body needs minerals such as magnesium and boron. It is then that the fire can be ignited in the belly through balancing pH. Fanning the flames for continuous burn comes from nourishing the microbiome-liver connection and providing it periods of rest. When all these steps are in place, then adding in nourishment through perfect-for-you food choices will fuel the flame to flood the cells with all they need to support an optimized, lean physique.

Remember, you are 100 percent unique and so is your body. Own it, and most importantly, serve it.

Continue to expand your connection with what your body needs, so that you can nourish it in the way it was designed to run. When you do, the results will speak for themselves. Nourishing your body daily in a customized manner will do the work on the inside, so you can rock your body on the outside!

REFERENCES

CHAPTER 4

- Li, H., Slone, J., Fei, L., & Huang, T. (2019). Mitochondrial DNA Variants and Common Diseases: A Mathematical Model for the Diversity of Age-Related mtDNA Mutations. *Cells, 8*(6), 608. https://doi.org/10.3390/cells8060608
- University of Hawai'i OER. (n.d.). Chapter 1: Nature of Science and Physics. In *College Physics Chapters 1-17*. essay. http://pressbooks-dev.oer.hawaii.edu/collegephysics/chapter/7-8-work-energy-and-power-in-humans/
- Wells, H. G. (2017). *Food of the gods*. Gollancz.

CHAPTER 5

- Escobedo, N., & Oliver, G. (2017). The lymphatic vasculature: Its role in adipose metabolism and obesity. *Cell metabolism*, *26*(4), 598–609. https://doi.org/10.1016/j.cmet.2017.07.020
- Grandjean, P., & Landrigan, P. J. (2014). Neurobehavioural effects of developmental toxicity. *The Lancet Neurology*, *13*(3), 330–338. https://doi.org/10.1016/s1474-4422(13)70278-3
- Haas, E. M., & Levin, B. (2006). Staying healthy with nutrition: The complete guide to diet and nutritional medicine. Berkeley: Celestial Arts.
- Ji, S. (2015, July 3). Why There is No Such Thing as 'Safe' Tap Water. GreenMedInfo. https://www.greenmedinfo.com/blog/why-there-no-such-thing-safe-tap-water
- Peckham, S., & Awofeso, N. (2014). Water fluoridation: a critical review of the physiological effects of ingested fluoride as a public health intervention. *The Scientific World Journal*, *2014*, 293019. https://doi.org/10.1155/2014/293019
- UPMC. (2016, March). Hydration: The Importance of Replacing Sweat Losses. [PDF].

CHAPTER 6

- Bakke D, Chatterjee I, Agrawal A, Dai Y, & Sun J. Regulation of Microbiota by Vitamin D Receptor: A Nuclear Weapon in Metabolic Diseases. Nucl Receptor Res. 2018;5:101377. doi: 10.11131/2018/101377. Epub 2018 Aug 9. PMID: 30828578; PMCID: PMC6392192.
- Bivona, G., Agnello, L., Bellia, C., Iacolino, G., Scazzone, C., Lo Sasso, B., & Ciaccio, M. (2019). Non-Skeletal Activities of Vitamin D: From Physiology to Brain Pathology. *Medicina, 55*(7), 341. https://doi.org/10.3390/medicina55070341
- Gaby, A. (2017). *Nutritional medicine* (2nd ed.). Concord, NH: Fritz Perlberg Publishing.
- Gittleman, A. L., & Burke, V. (2020). *Radical metabolism: a powerful new plan to blast fat and reignite your energy in just 21 days.* New York, NY: Hachette Go.
- Hollis, B. W., & Wagner, C. L. (2017). New insights into the vitamin D requirements during pregnancy. *Bone Research, 5*(1). https://doi.org/10.1038/boneres.2017.30
- Keegan, R. J., Lu, Z., Bogusz, J. M., Williams, J. E., & Holick, M. F. (2013). Photobiology of vitamin D in mushrooms and its bioavailability in humans. *Dermato-endocrinology, 5*(1), 165–176. https://doi.org/10.4161/derm.23321
- Liu, N. Q., Kaplan, A. T., Lagishetty, V., Ouyang, Y. B., Ouyang, Y., Simmons, C. F., Equils, O., & Hewison, M.

(2011). Vitamin D and the Regulation of Placental Inflammation. *The Journal of Immunology, 186*(10), 5968–5974. https://doi.org/10.4049/jimmunol.1003332

- Quig, D. & Maggiore, J. (2013). The standard: Does your vitamin d test measure up? Doctor's Data- Science & Insight. [PDF].
- Pludowski, P., Holick, M. F., Pilz, S., Wagner, C. L., Hollis, B. W., Grant, W. B., . . . Soni, M. (2013). Vitamin D effects on musculoskeletal health, immunity, autoimmunity, cardiovascular disease, cancer, fertility, pregnancy, dementia and mortality—A review of recent evidence. *Autoimmunity Reviews, 12*(10), 976-989. doi:10.1016/j.autrev.2013.02.004
- Proal, A. D., Albert, P. J., Blaney, G. P., Lindseth, I. A., Benediktsson, C., & Marshall, T. G. (2011). Immunostimulation in the era of the metagenome. *Cellular & Molecular Immunology*, 8(3), 213-225. doi:10.1038/cmi.2010.77
- Ruemmele, F. M., & Garnier-Lengliné, H. (2012). Why Are Genetics Important for Nutrition? Lessons from Epigenetic Research. *Annals of Nutrition and Metabolism,* 60(S3), 38-43. doi:10.1159/000337363
- Wholistic Methylation. (2020.). Methylation Testing. http://www.wholisticmethylation.com/
- Sitrin, M. (1978). Vitamin D Deficiency and Bone Disease in Gastrointestinal Disorders. *Archives of Internal Medicine, 138*(Suppl_5), 886. https://doi.org/10.1001/archinte.1978.03630300054011

- Sun, J. (2017). Vitamin D/Vitamin D Receptor regulation of microbiome in inflammation and obesity. *Journal of Metabolic Syndrome*, *06*(03). https://doi.org/10.4172/2167-0943-c1-004
- Williamson, C., & Pizano, J. (2018). Nutritional Genomics Institute. Retrieved from http://www.nutritionalgenomicsinstitute.com/

CHAPTER 7

- Burton, S. (2014, March 21). *How many bowls of Anything does it take to get the nutrition of Spirulina?* Down to Earth Organic and Natural. https://www.downtoearth.org/articles/2009-03/65/how-many-bowls-anything-does-it-take-to-get-nutrition-spirulina.
- Gaby, A. (2017). *Nutritional medicine* (2nd ed.). Concord, NH: Fritz Perlberg Publishing.
- Goodman, D. (2014). Magnificent magnesium: Your essential key to a healthy heart and more. Garden City Park, NY: Square One Publishers.
- Hoffer, A. & Walker, M. (1998). *Putting It All Together: the new orthomolecular nutrition*. Mcgraw-Hill Education.
- Ji, S. (2015, July 3). *The Amazing Healing Properties of Apples*. GreenMedInfo. https://www.greenmedinfo.com/blog/why-apple-one-worlds-most-healing-superfoods

- Pizzorno L. (2015). Nothing Boring About Boron. *Integrative medicine (Encinitas, Calif.), 14*(4), 35–48.
- Schnurr, K. (2016, January 11). *Magnesium and Hormonal Balance.* Kristin Schnurr, N.D. http://www.drkristinschnurr.com/blog/2016/1/10/magnesium-and-hormonal-balance-1
- Stargrove, M. B. (2008). Herb, nutrient, and drug interactions: Clinical implications and therapeutic strategies. St. Louis, Mo: Mosby/Elsevier.
- Weatherby, D. (2004). Signs and symptoms analysis from a functional perspective. Jacksonville: Bear Mountain Publishing.

CHAPTER 8

- Lipski, E. (2020). *Digestive wellness: strengthen the immune system and prevent disease through healthy digestion.* New York, NY: McGraw-Hill.
- McCoskey, L. (2019, November 23). *Powerful Powder: UTI relief and other magical things to know about Cream of Tartar.* Living Well Spine Center. http://livingwellspinecenter.com/blog/powerful-powder-uti-relief-and-other-magical-things-to-know-about-cream-of-tartar
- Weatherby, D. (2004). Signs and symptoms analysis from a functional perspective. Jacksonville: Bear Mountain Publishing.
- Wright, J. V., & Lenard, L. (2001). Why stomach acid

is good for you: natural relief from heartburn, indigestion, reflux, and Gerd. M. Evans.
• Young, R. O., & Young, S. R. (2010). *The pH miracle: balance your diet, reclaim your health*. New York, NY: Grand Central Life & Style.

CHAPTER 9

• Araujo Martins, A. M. (2016). Leptin Levels and its Relationship to Liver Dysfunctional Diseases and Hepatocellular Carcinoma. *Journal of Gastroenterology, Pancreatology & Liver Disorders, 3*(5), 01–07. https://doi.org/10.15226/2374-815x/3/5/00171
• Bajzer, M., & Seeley, R. J. (2006). Obesity and gut flora. *Nature, 444*(7122), 1009–1010. https://doi.org/10.1038/4441009a
• Bays, J. C. (2017). Mindful eating. A guide to rediscovering a joyful and healthy relationship with food, Revised Edition. Boston, MA: Shambhala.
• Bengmark S. "Nutrition of the Critically Ill: A 21st Century Perspective" *Nutrients* 2013, 5, 162-207
• Blatchford P, et al. *International Journal of Probiotics and Prebiotics* Vol. 8, No. 4, pp. 109-132, 2013.
• Brimeyer, T. (2017, November 30). Hypothyroidism and Adrenal Fatigue: Why Treating Your Adrenals Is Ruining Your Thyroid. Retrieved from http://www.forefronthealth.com/hypothyroidism-and-adrenal-fatigue/

- Conlon, M., & Bird, A. (2014). The Impact of Diet and Lifestyle on Gut Microbiota and Human Health. Nutrients, 7(1), 17-44. doi:10.3390/nu7010017
- De Alba, I. (n.d.) Microbiome: A new dimension in medicine. [PDF].
- Gittleman, A. L., & Burke, V. (2020). Radical metabolism: a powerful new plan to blast fat and reignite your energy in just 21 days. Hachette Go.
- Greenlaw, P., Ruggiero, M., & Greenlaw, D. (2015). *Your third brain: The revolutionary new discovery to achieve optimum health.* Centennial, CO: Extraordinary Wellness Publishing.
- Haas, E. M., & Levin, B. (2006). Staying healthy with Nutrition: the complete guide to diet and nutritional medicine. Berkeley: Celestial arts.
- Habib, N. (2019). *Activate your vagus nerve: unleash your body's natural ability to heal gut sensitivities, inflammation, brain fog, autoimmunity, anxiety, depression.* Berkeley, CA: Ulysses Press.
- Kalra, S. P. (1999). Interacting Appetite-Regulating Pathways in the Hypothalamic Regulation of Body Weight. *Endocrine Reviews, 20*(1), 68–100. https://doi.org/10.1210/er.20.1.68
- Lewitt, M. S., & Brismar, K. (2002). Gender difference in the leptin response to feeding in peroxisome-proliferator-activated receptor-alpha knockout mice. *International Journal of Obesity, 26*(10), 1296–1300. https://doi.org/10.1038/sj.ijo.0802135

- Lipski, E. (2012). Digestive wellness: strengthen the immune system and prevent disease through healthy digestion. New York, NY: McGraw-Hill.
- Liu, G. (2017). *Oxytocin: The hormone of mastery* [PDF]. Retrieved from The Gut Institute Website: https://thegutinstitute.com/wp-content/uploads/2017/09/pfx17-ho-oxytocin-hormone-of-mastery-compressed.pdf
- Lustig, R. (2013, Oct 18). University of California Television (UCTV). Fat Chance: Fructose 2.0. [Video File]. Retrieved from https://www.youtube.com/watch?v=ceFyF9px20Y
- Meyers, A. (2017, March). The Adrenal-Thyroid Connection. Retrieved from https://www.amymyersmd.com/2017/03/adrenal-thyroid-connection/
- McCoskey, L. (2019, November 23). *Powerful Powder: UTI relief and other magical things to know about Cream of Tartar*. Living Well Spine Center. http://livingwellspinecenter.com/blog/powerful-powder-uti-relief-and-other-magical-things-to-know-about-cream-of-tartar
- Neuman, H., Debelius, J. W., Knight, R., & Koren, O. (2015). Microbial endocrinology: the interplay between the microbiota and the endocrine system. *FEMS microbiology reviews*, 39(4), 509-521.
- Pert, C. B. (2003). Molecules of emotion: why you feel the way you feel. Scribner.
- Rignola, J. (Interviewer) & Kiran Krishnan, K. (Interviewee). (1999). Rebel Health Tribe: Metabolism,

weight loss, and out of control food cravings [Interview transcript]. Retrieved from Johnson Space Center Oral Histories Project website: ttps://rebelhealthtribe.com/course/microbiome-series-2-0/

- Temple, J. L., Cordero, P., Li, J., Nguyen, V., & Oben, J. A. (2016). A Guide to Non-Alcoholic Fatty Liver Disease in Childhood and Adolescence. International Journal Of Molecular Sciences, 17(6), doi:10.3390/ijms17060947
- Thomas, F., Hehemann, J. H., Rebuffet, E., Czjzek, M., & Michel, G. (2011). Environmental and gut bacteroidetes: the food connection. *Frontiers in microbiology*, 2, 93. https://doi.org/10.3389/fmicb.2011.00093
- Varian, B. J., Poutahidis, T., DiBenedictis, B. T., Levkovich, T., Ibrahim, Y., Didyk, E., Shikhman, L., Cheung, H. K., Hardas, A., Ricciardi, C. E., Kolandaivelu, K., Veenema, A. H., Alm, E. J., & Erdman, S. E. (2017). Microbial lysate upregulates host oxytocin. *Brain, Behavior, and Immunity*, *61*, 36–49. https://doi.org/10.1016/j.bbi.2016.11.002
- Verreth, W., De Keyzer, D., Pelat, M., Verhamme, P., Ganame, J., Bielicki, J. K., Mertens, A., Quarck, R., Benhabilès Nora, Marguerie Gérard, Mackness, B., Mackness, M., Ninio, E., Herregods, M.-C., Balligand, J.-L., & Holvoet, P. (2004). Weight Loss–Associated Induction of Peroxisome Proliferator–Activated Receptor-α and Peroxisome Proliferator–Activated Receptor-γ Correlate With Reduced Atherosclerosis and

Improved Cardiovascular Function in Obese Insulin-Resistant Mice. *Circulation, 110*(20), 3259–3269. https://doi.org/10.1161/01.cir.0000147614.85888.7a
• Welch, M. G., Anwar, M., Chang, C. Y., Gross, K. J., Ruggiero, D. A., & Gershon, M. D. (2010). Combined administration of secretin and oxytocin inhibits chronic colitis and associated activation of forebrain neurons. *Neurogastroenterology and Motility : The Official Journal of the European Gastrointestinal Motility Society, 22*(6), 654–e202.

CHAPTER 10

• Campbell, W. R. (2012) Vibrations: Vibrational Frequencies and Food. Returned Bliss. Retrieved from https://blissreturned.wordpress.com/2012/02/03/vibrations-vibrational-frequencies-and-food/amp/
• Dimauro S, Davidzon G. Mitochondrial DNA and disease. Ann Med. 2005;37(3):222-32.
• Hoffer, A. & Walker, M. (1998). *Putting It All Together: the new orthomolecular nutrition*. Mcgraw-Hill Education.
• Lipski, E. (2020). Digestive wellness: strengthen the immune system and prevent disease through healthy digestion. McGraw-Hill.
• Marchegiani, J. (2019, July 18). *Mitochondria 101: The Key to Longevity*. Austin Texas Functional Medicine and

Nutrition. https://justinhealth.com/mitochondria-101-the-key-to-longevity/

• McInnes, J. (2013). Mitochondrial-associated metabolic disorders: foundations, pathologies and recent progress. *Nutrition & Metabolism*, 10(1), 63. https://doi.org/10.1186/1743-7075-10-63

ABOUT THE AUTHOR

Arianne is a Certified Nutrition Specialist®, the founder of Aceso Nutrition, and the author of Skinny Genes: The Surprising Truth about Every Body's Capacity to Settle at a Natural Weight, Even When Diets Have Failed.

Arianne has a BS in Cellular Molecular Biology and Psychology and an MS in Nutrition and Integrative Health. She specializes in helping clients discover what is holding them back from achieving their perfect body through a body-centered approach to nutrition. She is passionate about getting people out of their head and into their body to figure out the body-perfect diet for their 100 percent unique self, so they can achieve and sustain their long-term nutrition goals.

Arianne is thrilled to bring her robust analytical, critical thinking, and strategy development skills from years serving in the United States Marine Corps intelligence community and as an independent strategic business consultant to her role as a nutritionist. Her history of solving hard problems through innovative approaches and keen observation has helped shape Arianne's model for her nutrition clients.

Arianne teaches clients how to unlock their skinny genes through the tailor-made diet program, seminars, retreats, books, speaking engagements, and precision nutrition one-on-one sessions.

She lives in Northern Virginia with her husband and two kiddos.

Website: www.theaceso.com
Email: ambozarth@theaceso.com

ABOUT DIFFERENCE PRESS

DIFFERENCE
P R E S S

Difference Press is the exclusive publishing arm of The Author Incubator, an educational company for entrepreneurs – including life coaches, healers, consultants, and community leaders – looking for a comprehensive solution to get their books written, published, and promoted. Its founder, Dr. Angela Lauria, has been bringing to life the literary ventures of hundreds of authors-in-transformation since 1994.

A boutique-style self-publishing service for clients of The Author Incubator, Difference Press boasts a fair and easy-to-understand profit structure, low-priced author copies, and author-friendly contract terms. Most importantly, all of our #incubatedauthors maintain ownership of their copyright at all times.

LET'S START A MOVEMENT WITH YOUR MESSAGE

In a market where hundreds of thousands of books are published every year and are never heard from again, The Author Incubator is different. Not only do all Difference Press books reach Amazon bestseller status, but all of our authors are actively changing lives and making a difference.

Since launching in 2013, we've served over 500 authors who came to us with an idea for a book and were able to write it and get it self-published in less than 6 months. In addition, more than 100 of those books were picked up by traditional publishers and are now available in bookstores. We do this by selecting the highest quality and highest potential applicants for our future programs.

Our program doesn't only teach you how to write a book – our team of coaches, developmental editors, copy editors, art directors, and marketing experts incubate you from having a book idea to being a published, bestselling author, ensuring that the book you create can actually make a difference in the world. Then we give you the training you need to use your book to make the difference in the world, or to create a business out of serving your readers.

ARE YOU READY TO MAKE A DIFFERENCE?

You've seen other people make a difference with a book. Now it's your turn. If you are ready to stop watching and start taking massive action, go to http://theauthorincubator.com/apply/.

"Yes, I'm ready!"

OTHER BOOKS BY DIFFERENCE PRESS

Beyond the Why of Loss: A Brave New Way to Move Forward by Elaine Alpert, M.Ed

Girl, It's Time to Move On: 5 Practices for Healing after a Breakup or Divorce by Daye Ambersley

Out of the Man Cave, Into the Heart of the Goddess: The Modern Man's Guide to Conscious Love, Intimacy, and the Awakened Woman by Lord Coltrane

Academy of Eternity: Unlock the Full Potential of Your Heart-Mind, Now and Forever by Erika Flint, Sarah Solstice, and the Babyji, with Sam Tullman

A Guide to Sensing and Feeling Energy: Discover Your Unique Abilities by Heidi Henyon

The Food Solution: Eating for Today to Save Tomorrow by Dr. Gundula Rhoades

When You've Outgrown Your Life: How to Avoid Losing Precious Time Because You Fear Losing it All by Rita Sampaio, PhD

Client Magnet: The Coaches Guide to Attract Ideal Clients through Spiritual Awareness by Cheryl Stelte

Releasing Self-Doubt: A Holistic Guide to Letting Go of What Other People Think and Finally, Fully Believing in Yourself by Joy Stone

THANK YOU

Thank you so much for reading *Skinny Genes: The Surprising Truth about Every Body's Capacity to Settle at a Natural Weight, Even When Diets Have Failed*. If you have made it this far, then I know you are committed to designing a diet that is 100 percent made for your body, so you can wear your skinny jeans each and every day!

To keep you inspired on your journey, I've created a video series that provides tips and tricks to designing each component of your customized diet. Spend a week with me exploring! Each day you will receive a "Skinny Genes: Diet by Design" video by email. Sign up at www.theaceso.com.